Philippe Bradac

Interactions of cadmium with periphyton in natural freshwaters

Philippe Bradac

Interactions of cadmium with periphyton in natural freshwaters

Experimental studies on the biogeochemistry and
bioavailability of metals in aquatic systems

Südwestdeutscher Verlag für Hochschulschriften

Impressum / Imprint
Bibliografische Information der Deutschen Nationalbibliothek: Die Deutsche Nationalbibliothek verzeichnet diese Publikation in der Deutschen Nationalbibliografie; detaillierte bibliografische Daten sind im Internet über http://dnb.d-nb.de abrufbar.
Alle in diesem Buch genannten Marken und Produktnamen unterliegen warenzeichen-, marken- oder patentrechtlichem Schutz bzw. sind Warenzeichen oder eingetragene Warenzeichen der jeweiligen Inhaber. Die Wiedergabe von Marken, Produktnamen, Gebrauchsnamen, Handelsnamen, Warenbezeichnungen u.s.w. in diesem Werk berechtigt auch ohne besondere Kennzeichnung nicht zu der Annahme, dass solche Namen im Sinne der Warenzeichen- und Markenschutzgesetzgebung als frei zu betrachten wären und daher von jedermann benutzt werden dürften.

Bibliographic information published by the Deutsche Nationalbibliothek: The Deutsche Nationalbibliothek lists this publication in the Deutsche Nationalbibliografie; detailed bibliographic data are available in the Internet at http://dnb.d-nb.de.
Any brand names and product names mentioned in this book are subject to trademark, brand or patent protection and are trademarks or registered trademarks of their respective holders. The use of brand names, product names, common names, trade names, product descriptions etc. even without a particular marking in this work is in no way to be construed to mean that such names may be regarded as unrestricted in respect of trademark and brand protection legislation and could thus be used by anyone.

Verlag / Publisher:
Südwestdeutscher Verlag für Hochschulschriften
ist ein Imprint der / is a trademark of
OmniScriptum GmbH & Co. KG
Heinrich-Böcking-Str. 6-8, 66121 Saarbrücken, Deutschland / Germany
Email: info@svh-verlag.de

Herstellung: siehe letzte Seite /
Printed at: see last page
ISBN: 978-3-8381-1618-1

Zugl. / Approved by: Zurich, Swiss Federal Institute of Technology Zurich (ETH), Ph.D. Thesis, 2009

Copyright © 2010 OmniScriptum GmbH & Co. KG
Alle Rechte vorbehalten. / All rights reserved. Saarbrücken 2010

Table of content

SUMMARY 5

1. INTRODUCTION 11
   1.1. TRACE METALS IN THE AQUATIC ENVIRONMENT 13
   1.2. METAL SPECIATION 14
   1.3. DETERMINATION OF METAL SPECIATION 15
   1.4. PERIPHYTON 17
   1.5. METAL SPECIATION AND BIOAVAILABILITY 18
   1.6. INTRACELLULAR METAL DISTRIBUTION 20

2. KINETICS OF CADMIUM ACCUMULATION IN PERIPHYTON UNDER FRESHWATER CONDITIONS 23
   2.1. ABSTRACT 25
   2.2. INTRODUCTION 25
   2.3. MATERIALS AND METHODS 27
      2.3.1. Channels setup for periphyton colonisation 27
      2.3.2. Kinetic experiment 28
      2.3.3. Clean trace metal handling 29
      2.3.4. Water sampling for dissolved metal concentrations 29
      2.3.5. Preparation, sampling and processing of DGT devices 29
      2.3.6. Periphyton sampling and processing 30
      2.3.7. Metal analysis 31
      2.3.8. Water chemistry 32
      2.3.9. Modelling of intracellular cadmium uptake and release in periphyton 32
   2.4. RESULTS 34
      2.4.1. Dissolved and DGT-labile metal concentrations in water 34
      2.4.2. Periphyton characterization 36
      2.4.3. Kinetics of cadmium uptake and release in periphyton 37
      2.4.4. Linear and non-linear modelling of cadmium accumulation in periphyton 39
   2.5. DISCUSSION 41
      2.5.1. Metal speciation in channels 41
      2.5.2. Cadmium accumulation and kinetics in periphyton 41
      2.5.3. Competition of other metals for cadmium uptake 44
      2.5.4. Cadmium release 44
      2.5.5. Environmental relevance 45

3. **ACCUMULATION OF CADMIUM IN PERIPHYTON UNDER VARIOUS FRESHWATER SPECIATION CONDITIONS 47**
   - 3.1. ABSTRACT 49
   - 3.2. INTRODUCTION 49
   - 3.3. MATERIALS AND METHODS 50
     - 3.3.1. Periphyton colonization 50
     - 3.3.2. Experimental design... 51
     - 3.3.3. Clean trace metal handling.. 51
     - 3.3.4. Sampling.. 52
     - 3.3.5. NOM analysis and modelling of metal speciation 52
     - 3.3.6. Preparation and processing of DGT devices.............................. 53
     - 3.3.7. Periphyton processing.. 54
     - 3.3.8. Analytical methods 54
   - 3.4. RESULTS 55
     - 3.4.1. Dissolved and DGT-labile metal concentrations....................... 55
     - 3.4.2. NOM characterization 58
     - 3.4.3. Modelled metal species 58
     - 3.4.4. Periphyton characterisation 59
     - 3.4.5. Metal accumulation and speciation 59
   - 3.5. DISCUSSION 61
     - 3.5.1. Metal speciation in channels 61
     - 3.5.2. Accumulation and speciation 63
     - 3.5.3. Environmental relevance 64

4. **CADMIUM SPECIATION AND ACCUMULATION IN PERIPHYTON IN A SMALL STREAM DURING RAIN EVENTS ... 65**
   - 4.1. ABSTRACT 67
   - 4.2. INTRODUCTION 67
   - 4.3. MATERIALS AND METHODS 69
     - 4.3.1. Site description... 69
     - 4.3.2. Periphyton colonization and translocation................................ 69
     - 4.3.3. Field study 70
     - 4.3.4. Clean trace metal handling.. 71
     - 4.3.5. Water sampling for dissolved metal concentrations.................. 71
     - 4.3.6. NOM analysis and modelling of metal speciation 71
     - 4.3.7. Preparation, sampling and processing of DGT devices............ 72
     - 4.3.8. Periphyton sampling and processing.. 73
     - 4.3.9. Metal analysis 73
     - 4.3.10. Water chemistry 74

4.4.	RESULTS	74
4.4.1.	Dissolved metal concentrations	74
4.4.2.	DGT-labile metal concentrations	78
4.4.3.	NOM characterisation	79
4.4.4.	Modelled metal species	80
4.4.5.	Periphyton characterisation	83
4.4.6.	Metal accumulation in periphyton	83
4.5.	DISCUSSION	85
4.5.1.	Metal speciation	85
4.5.2.	Metal accumulation in periphyton	87
4.5.3.	Environmental implications	89

OUTLOOK ... 91

REFERENCES ... 95

APPENDIX ... 109

SUPPORTING INFORMATION TO CHAPTER 4 ... 111
    Modelled metal species (S1) ... 111
    Metal content in periphyton (S2) ... 117

PICTURES ... 119
    Kinetics of cadmium accumulation in periphyton under freshwater conditions ... 119
    Accumulation of cadmium in periphyton under various freshwater speciation conditions ... 121
    Cadmium speciation and accumulation in periphyton in a small stream during rain events ... 123
    Periphyton slides ... 125

ACKNOWLEDGEMENTS ... 127

Summary

Summary

Cadmium is a non-essential trace metal, with elevated concentrations in many aquatic systems. The bioavailability of cadmium as well as of other non-essential and essential metals depends not only on their concentrations but is also strongly dependent upon chemical speciation. Models for the bioavailability of metals have been developed in order to predict uptake and effects. Equilibrium-based models, i.e. the free ion activity model (FIAM) and the biotic ligand model (BLM) have been confirmed in various laboratory studies with algae in defined media. Exceptions which have been found to these models were due to the uptake of specific metal-ligand complexes and the control of bioaccumulation by labile metal species. The latter case highlights the importance of including the dynamic aspects of metal complexes in non-equilibrium based models, by considering diffusion and uptake fluxes of such complexes.

Only a few studies have related metal accumulation in algae to metal speciation under natural conditions. Periphyton is the natural algal biofilm in surface waters and as the predominant primary producer of ecological importance. Since different natural and anthropogenic ligands present in aquatic systems influence metal speciation and periphyton is a complex three-dimensional community of various algal species imbedded in an organic matrix, differences may be expected between cadmium accumulation by periphyton in natural waters and single algal species in defined media.

The aim of the thesis was to provide missing information about the interactions of cadmium with periphyton in natural freshwaters. Therefore experiments were conducted under semi-controlled conditions in artificial channels supplied with natural freshwater, which allowed cadmium concentrations and speciation to be controlled. To validate the results from these experiments and to study the interactions of cadmium with periphyton under natural conditions a field study was performed.

Summary

The first chapter provides information about the geochemistry of metals, in particular which factors influence the speciation of metals and how it can be measured. Examples are given how metal speciation influences the bioavailability of metals in algae and in periphyton. Furthermore, information on the ecology and structure of periphyton is given and the complexity of processes influencing speciation, bioavailability and finally uptake of metals by algae is described. Finally, the processes by which algae control intracellular metal concentrations, their release and effects of metals on the algal physiology are summarized.

In a first study (Chapter 2) the kinetics of cadmium uptake and release in periphyton were investigated at environmentally relevant cadmium concentrations in artificial flow-through channels over a period of 26.4 hours. Speciation of cadmium and other metals was determined using diffusion gradient in thin-films (DGT). Two uptake phases, a fast initial phase and a slower subsequent phase that reached steady state by the end of cadmium exposure were observed. Uptake rates and bioconcentration factors were obtained. A one compartment model was used to fit the uptake during the cadmium exposure from which uptake and clearance rate constants were obtained The competition of other metals and the influence of the periphyton structure on cadmium uptake are discussed. Compared to experiments with single algal species in chemically controlled media, the kinetics of cadmium uptake in periphyton were much slower, which might be due to a slower diffusion of cadmium through the extracellular matrix of periphyton. Cadmium concentrations in periphyton decreased slowly after cadmium addition was stopped, probably because of a strong complexation by intracellular ligands.

In chapter 3 the relationship between cadmium accumulation in periphyton and cadmium speciation in water was investigated using artificial recirculating channels and natural freshwater with different cadmium speciation conditions. They were achieved by adding an artificial organic ligand (nitrilotriacetate, NTA) to two different cadmium concentrations. The ligand increased DGT-labile cadmium concentrations, which were dominated by Cd-NTA complexes and simultaneously decreased free cadmium concentrations. In contrast to the prediction of equilibrium based models (FIAM, BLM) accumulation did not depend on the free cadmium concentrations, but on the diffusion of DGT-labile complexes to the algae and highlighted the importance of considering dynamic species in metal accumulation.

In chapter 4 a field study in the small stream Altbach (Switzerland) was conducted in order to validate the results from these semi-controlled experiments and to investigate the relationships between speciation and accumulation under natural conditions. The changes of metal content in periphyton were investigated as a function of dynamic variations of cadmium speciation in water during rain events. Periphyton responded rapidly and sensitively to variations of Cd concentrations in water, despite an excess of Zn and Mn. Cadmium accumulation was related to labile cadmium concentrations which were measured *in situ* with the technique of diffusion gradients in thin-films (DGT) and to modelled labile and free cadmium concentrations. Dissolved Cd concentrations increased from 0.17 nM to concentrations between 0.27 and 0.36 nM and modelled free Cd concentrations from 0.10 nM to concentrations between 0.12 and 0.14 nM. DGT-labile Cd concentrations did not change during the rain events and were between 70 and 97% of total dissolved Cd. DGT-labile cadmium concentrations were compared with modelled labile concentrations, by including diffusion coefficients of humic substances. They agreed well with modelled labile Cd concentrations showing that cadmium was mostly present in labile and free metal ion form. Therefore a high percentage of total dissolved cadmium was available for uptake by periphyton. The dynamics of accumulation were different during the rain events, which can be explained with the interactions of other metals and intracellular metal regulation mechanisms.

The results from these studies show that periphyton sensitively responds to changes of environmentally relevant Cd concentrations in water and that differences were observed to algae experiments in defined media, probably due to the structure of periphyton. Labile metal complexes with organic ligands might control

Summary

bioaccumulation in periphyton in natural freshwaters. Thus, dynamic metal species should be considered in models predicting bioavailability.

Chapter 1

Introduction

1.1. Trace metals in the aquatic environment

Trace metals of concern in aquatic environments include both essential (e.g. Cu, Zn, Mn, Fe) and non-essential elements (e.g. Cd, Pb, Hg) [1]. Essential trace metals are important as cofactors for metalloenzymes and proteins, for the maintenance of conformations and tertiary structures of proteins and for the catalysis of redox reactions and electron transport [2]. Human activities have increased concentrations of essential as well as non-essential metals (e.g. Cd, Pb, Hg) over background levels in many aquatic systems [3-6]. Organisms possess a specific physiological concentration range for essential trace metals, within which vital processes are maintained. Concentrations under or above this range will lead to deficiencies or toxicity. In contrast non-essential metals can already be toxic at low concentrations, and toxicity increases with increasing concentrations [7]. Metal toxicity arises from the displacement of essential metals in metalloenzymes, the blocking of functional groups or conformation modification of biomolecules [8].

Cadmium is considered as a priority pollutant in freshwaters [9]. It is introduced into the environment from several direct or indirect anthropogenic sources. It is used in industry as a protective plating on steel, a stabilizer for PVC, pigments in plastics and glasses, an electrode material in nickel-cadmium batteries and as a component of various alloys. Cadmium is released into the environment during their production and combustion in refuse incineration plants. Other anthropogenic sources are phosphate fertilizers, mining, fossil fuel combustion and sewage sludge incineration [10].

Background concentrations of dissolved Cd in river water are between 0.03 and 0.2 nM [11-13]. But concentrations can be much higher at sites affected by metal pollution, e.g. 438 nM [3] downstream of coal minings and Zn ore treatment and between 4.2 µM [4] and 6.8 µM [5] downstream of metallurgical factories. The bioavailability and hence uptake and effects of elevated Cd concentrations to aquatic organisms depends not only on its concentration but is strongly influenced by its chemical speciation in water [7, 14].

Chapter 1

1.2. Metal speciation

Since natural freshwaters are temporally and spatially heterogeneous environments in terms of physical properties and chemical and biological composition, metals occur in a variety of chemical forms or species.

Within the dissolved phase metals are present either as free aqua ions or bound to inorganic and organic ligands. Inorganic ligands comprise for example hydroxide, carbonate, hydrogen carbonate and chloride [15, 16]. Natural organic ligands comprise a wide range of low and high molecular weight compounds. Small organic molecules include carboxylic acids (oxalate, acetate, malonate and citrate), amino acids, phenols and catechols. A very important part of larger organic ligands belongs to the class of humic and fulvic acids [17]. These polymeric compounds possess heterogeneous binding sites, mostly phenolic and carboxylic groups and in small amounts N- and S-containing functional groups [16]. Exudates of bacteria, algae and fungi (siderophores, phytochelatins) also contribute to the organic ligands [18, 19]. In addition to these natural organic ligands, synthetic organic ligands like EDTA and NTA are introduced by anthropogenic activities into the aquatic environment [20].

Metals may also be present in particulate form [21, 22]. Colloidal particles may consist of organic (e.g. humic acids) and inorganic (e.g. metal oxides, clay minerals) components. The particulate matter consists of oxides, hydroxides, carbonates, sulfides (anoxic conditions), of silicates and clay minerals, organic debris and microorganisms (algae, bacteria) [16].

As a consequence metal species with different complexing properties, with respect to kinetic lability and thermodynamic stability are present in natural waters [16, 21]. Kinetic lability refers to the dissociation rate of metals from their complexes and allows a distinction between labile and inert complexes. Labile complexes have lower stability constants and higher dissociation rates than inert complexes [21]. The lability of complexes can only be defined with respect to a given time scale (e.g. corresponding to a certain analytical technique).

1.3. Determination of metal speciation

Metal speciation in water can be either measured with different speciation techniques or estimated with chemical equilibrium models. Each speciation technique measures only a certain proportion of the total metal species present in water [23]. Some techniques measure only free metal ion concentrations, whereas others measure dynamic metal complexes or both (Table 1.1). Dynamic metal complexes are both mobile and labile, i.e. they have to diffuse through a gel to the sensor and dissociate within the time required to diffuse through the gel [23-25]. Speciation techniques can be used either directly in the field (*in situ*) or on discrete water samples in the laboratory (*ex situ*). *Ex situ* techniques have disadvantages related to 1) possible contamination during sampling, storage and measurement, 2) changes in metal speciation prior to the measurement and 3) adsorption of metals to the container walls [26].

Some available chemical speciation programs are ECOSAT (Wageningen University, The Netherlands), WHAM 6 [27, 28] and Visual MINTEQ [29]. They use different models for the complexation of metals by humic and fulvic substances, specifically the non-ideal competitive adsorption (NICA)-Donnan model [30], Model VI (WHAM-Windermere Humic Aqueous Model) [27, 28] and the Stockholm humic model (SHM) [31]. The accuracy of the models depends on the input parameters, i.e. concentrations and stability constants of species. Dissolved organic carbon is normally assumed to consist only of humic and fulvic acids and is usually divided by assumptions into certain ratios of these acids. A novel technique for the fractionation of dissolved organic carbon, size-exclusion chromatography coupled to on-line, high sensitivity organic carbon detection (LC-OCD) [32-34], allows the characterization of humic substances with respect to their aromaticity and molecular weights. Other organic compounds are also determined with this technique and can occur in similar concentrations as humic substances in water.

Chapter 1

Speciation technique	Measured metal species	in situ	ex situ	Dynamic technique[I]	Equilibrium-based technique[II]	Typical analysis time [s][9]
Flow-through permeation liquid membrane (FTPLM) [1]	free ions / dynamic		X	X	X	10^2-$10^{3\ \text{I})}$ / $10^{4\ \text{II})}$
Stripping chronopotentiometry (SCP) [2]	dynamic		X	X		10^2-10^3
Competitive ligand-exchange/stripping voltammetry (CLE-SV) [3]	free ions		X		X	10^2-10^3
Diffusion gradients in thin films (DGT) [4]	dynamic	X		X		10^3-10^5
Gel integrated microelectrode (GIME) [5]	dynamic	X		X		10^2-10^3
Donnan membrane technique (DMT) [6]	free ions	X		X	X	10^5
Hollow fiber permeation liquid membrane (HFPLM) [7]	free ions / dynamic	X		X	X	10^2-$10^{3\ \text{I})}$ / $10^{4\ \text{II})}$
Ion selective electrode (ISE) [8]	free ions	X	X		X	1-10

Table 1.1: Summary of currently available speciation techniques. 1) [35], 2) [36, 37], 3) [22], 4) [38-40], 5) [41]. It can be used in combination with a voltammetric in situ profiling systems or a voltammetric in-line analyzer (GIME-VIP or GIMEVIA-FIELD [42], 6) [43, 44], 7) [45], 8) [46]. 9) [23]. The ion selective electrode is the only technique, which is suitable only for relatively high free metal concentrations.

Speciation programs are useful for media with defined chemical composition, but they are based on the assumption of equilibrium, which is not always valid for natural systems. In addition, many components are unknown in natural waters and undefined organic ligands are present as well as humic and fulvic acids. Thus, such calculations must be compared with results obtained by speciation techniques.

1.4. Periphyton

Periphyton is the natural assemblage of autotrophic and heterotrophic organisms (algae, bacteria, fungi) [47], which grows on various substrates in the photic zones of streams, rivers, lakes and estuarine and marine ecosystems. In contrast to free floating algae, the organisms of periphyton are embedded in a heterogeneous matrix, which is secreted by both prokaryotic (bacteria, Archaea) and eukaryotic (algae, fungi) organisms [48]. The matrix consists of extracellular carbohydrates which can be broadly grouped into low molecular weight compounds (e.g. sugars, glycolates) and larger heteropolymeric substances (EPS = extracellular polymeric substances, e.g. polysaccharides) [49]. Other components are proteins, nucleic acids and (phospho) lipids [50]. The matrix is responsible for the adhesion to the substrate in water, for the cohesion of the community [48] and can absorb metals [51]. It has been shown that exudates of algae [52] and bacteria [53] are able to bind metals and that phytoplankton increased the production of extracellular polysaccharides with the exposure to metals [54]. The matrix therefore plays an important role in influencing the bioavailability of metals.

Periphyton is the most important primary producer in running waters [55]. Due to its settlement and structure it retains organic carbon [56] and inorganic nutrients [57, 58] and provides habitat and shelter for aquatic organisms and a food source for grazers and herbivores.

Periphyton accumulates cadmium and other metals effectively from water. Cadmium concentrations in periphyton from unpolluted sites are below 1 nmol Cd g dw^{-1} [13], but they are higher at moderately polluted sites (11.9 to 128 nmol Cd g dw^{-1} [5, 59]) and even much higher at sites impacted by metal pollution (16 and 167 µmol Cd g dw^{-1} [3-5]). Elevated Cd concentrations in water have been shown to change biomass related parameters (chlorophyll a, dry weight) [60, 61], settlement, development and species composition of periphyton [62] and there is also evidence of Cd biomagnification among trophic levels [63].

Chapter 1

1.5. Metal speciation and bioavailability

Metal internalization by algae generally includes three steps [64, 65] (Figure 1.1). First, the free metal ion or metal complex diffuses from the bulk solution through the diffusion layer and the cell wall to the plasma membrane. In periphytic algae the diffusion is further affected by the matrix. At the plasma membrane the free metal ion binds following chemical equilibrium to specific transport sites.

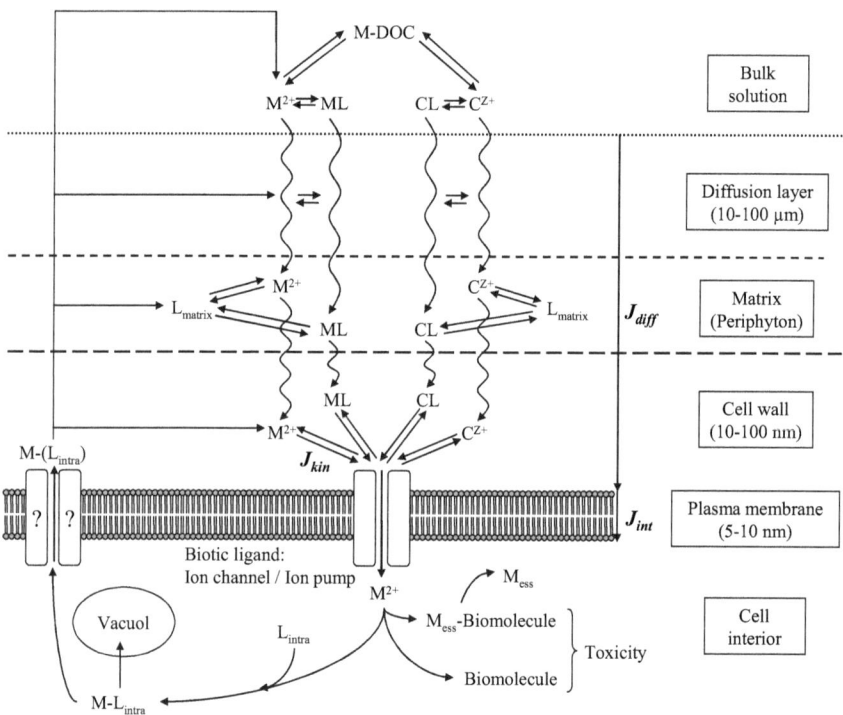

Figure 1.1: A summary of metal-algae interactions. The matrix is an exclusive feature of periphyton. Abbreviations: Dissolved organic carbon (DOC), inorganic ligands (L), free metal ions (M^{2+}), competing cations (C^{z+}), intracellular ligands (L_{intra}), essential trace metal (M_{ess}), matrix ligands (L_{matrix}), diffusive flux (J_{diff}), chemical dissociation flux (J_{kin}), internalization flux (J_{int}).

This adsorption is very fast, followed by a slower uptake across the plasma membrane. Transmembrane proteins involved in this uptake can be either ion channels, which facilitate the diffusion along an electrochemical gradient or ion transporters, which require energy. Intracellular metal concentrations are regulated by various mechanisms (see Chapter 1.6).

Based on early laboratory studies with algae in well defined culture media it became clear that bioavailability does not depend on the total dissolved metal concentration but on the speciation of the metal [66-69]. On that basis models have been developed, which relate uptake in aquatic organisms to metal speciation in water.

The free ion activity model (FIAM) predicts that biological responses are related to the activity of the free metal ion concentration [14, 70]. This model has been confirmed for various metals and algal species in laboratory studies [71-74]. However, since the development of the FIAM limitations and exceptions have been found. These have been explained by the formation of specific metal-ligand complexes, that were internalized by transporters other than metal cation transporters or by passive diffusion. An increased accumulation of silver was shown in *Chlamydomonas reinhardtii* in the presence of thiosulfate [75]. It was concluded that these metal-inorganic complexes were accidentally transported across the plasma membrane via sulphate/thiosulfate transport systems. A similar effect was observed with small metal-organic complexes in a study with *Selenastrum capricornutum* [76, 77]. The intracellular uptake of cadmium increased in the presence of the low molecular weight metabolite citrate, which was explained by an accidental transport of cadmium-citrate complexes across the plasma membrane. An uptake of lipophilic metal-organic complexes by passive diffusion across the plasma membrane was demonstrated for copper, cadmium and lead in *Thalassiosira weissflogii* [78] and for copper in five phytoplankton species [79]. Such lipophilic dissolved and particulate metal species were identified in river water [80].

Water chemistry and composition is much more complex in natural waters than in synthetic media. The biotic ligand model (BLM) [81], which was derived from the FIAM, takes other cations and natural ligands into account. Cations (metals, H^+, Ca^{2+}) can compete with metals of interest for uptake sites (biotic ligand) and natural organic ligands (DOC) can alter the speciation of both cations and metals in water (Figure 1.1).

Both FIAM and BLM are equilibrium-based models that neglect the dynamic features of metal complexes. They assume that 1) equilibrium is attained between the free metal ion in the bulk solution and that adsorbed to the biotic ligand and 2) the diffusion flux of metal species from the bulk solution to the plasma membrane (J_{diff}) is faster than the internalization flux (J_{int}). In contrast, non-equilibrium based models consider diffusion and internalization fluxes [23, 82, 83]. Diffusion of free metal ions to the plasma membrane can be limiting if the diffusion flux (J_{diff}) is smaller than the internalization flux (J_{int}) (Figure 1.1). In that case metal uptake is apparently controlled by the concentrations of chemically labile species [82, 84]. This was confirmed for silver uptake in *Chlamydomonas reinhardtii* in the presence of chloride [85] and for copper uptake in periphyton at low environmental free copper concentrations [86]. In periphyton the diffusion of metal ions or metal complexes (J_{diff}) might be further limited by the matrix (Figure 1.1).

1.6. Intracellular metal distribution

Cellular metal ion concentrations need to be regulated in order to maintain the physiological needs of the cell and to prevent toxic effects. This is generally achieved by chelation, distribution, sequestration and release [8, 87]. Essential metal ions are bound by chaperons, which deliver them to specific organelles and metal-requiring proteins. To prevent toxic effects of both essential and non-essential metals, metal ions can be either sequestered directly into vacuoles or bound to chelators (phytochelatins, organic acids, amino acids), which can be also transported into vacuoles or released from the cell [18, 87, 88].

The toxicity of cadmium arises from its non-specific binding to physiologically important biomolecules (Figure 1.1). It can substitute for essential metal ions or directly bind to functional groups, reducing or suppressing biological activity. In algae, phytochelatins (PCs) play an important role in metal chelation and detoxification of elevated intracellular metal concentrations. PCs are small metal-binding polypeptides with the amino acid composition (γ-Glu-Cys)$_n$-Gly where n is between 2 and 11. They have a high affinity to various metals and form Me-PC$_n$ complexes due to their sulfhydryl and carboxylic acid groups. It has been shown that PC production is strongly induced in algae by cadmium even at low concentrations [89, 90]. In plants, Me-PCn complexes are sequestered in vacuoles [88]; in fission yeast, transport of such complexes into vacuoles is mediated by a protein (ATP-binding cassette) [91]. Accumulated cadmium has been demonstrated to be released from algae [74, 77] as well as periphyton [13, 92, 93]. The release of Cd as Cd-PC complexes has been demonstrated for a marine diatom [18], but Cd could be also released as the free cadmium ion (Figure 1.1). Released cadmium might then either interact with algal transport sites or diffuse back in the diffusion layer or the bulk solution. In case of periphyton the released Cd might also adsorb to ligands of the matrix.

The subcellular partitioning model (SPM) has been developed to predict metal toxicity to small aquatic organisms [94]. It assumes that toxicity depends on the distribution of the metal among different ligands and not on the total amount accumulated. The authors differentiate operationally between metals bound to sensitive ligands and those bound to detoxifying ligands. In this model, elevated metal concentrations are assumed to saturate detoxifying ligands, leading to a redistribution of metals to sensitive ligands with resulting toxicity [95]. Detoxification mechanisms and hence this distribution and effects of metals may be species specific. Indeed, the toxicity of Cd differs significantly among different species of marine phytoplankton [96].

Chapter 2

Kinetics of cadmium accumulation in periphyton under freshwater conditions

This chapter is published in *Environmental Toxicology and Chemistry, 2009, 28 (10): 2108–2116*

2.1. Abstract

The aim of the study was to investigate the kinetics of Cd accumulation (total and intracellular) in periphyton under freshwater conditions in a short-term microcosm experiment. Periphyton was precolonised in artificial flow-through channels supplied with natural freshwater and then exposed for 26.4 hours to nominal Cd concentrations of 5 and 20 nM added to natural freshwater. Labile Cd in water determined with diffusion gradients in thin films (DGT) was 60 to 69% of total dissolved Cd in the exposure channels and 11 % in the control channel. Intracellular Cd concentrations in periphyton increased rapidly and linearly during the first 71 min. Initial intracellular uptake rates were 0.05 and 0.18 nmol Cd/g dw X min in the 5 nM and the 20 nM exposures, respectively. The subsequent intracellular uptake was slower, reaching steady state at the end of Cd exposure. Cadmium uptake kinetics were slower when compared to experiments with planktonic algal cultures probably due to diffusion limitations. Intracellular Cd uptake during the entire exposure was modelled with a non-linear, one compartment model, from which uptake and clearance rate constants as well as bioconcentration factors were obtained. The release of Cd from periphyton after end of Cd exposure was slow, when compared to the initial uptake rates.

2.2. Introduction

Cadmium is a very toxic, non-essential trace metal, which has been designated a priority pollutant. It is introduced into the environment by several anthropogenic sources, e.g. phosphate fertilizers, mining or fossil fuel combustion [10]. Concentrations of dissolved Cd in pristine river water are between 0.03 and 0.2 nM [11, 12], but can be much higher near sources of metal pollution, e.g. 438 nM [3] downstream of coal mining and Zn ore treatment and between 4.2 µM [4] and 6.8 µM [5] downstream of metallurgic factories.

Bioavailability and hence uptake and effects of trace metals on aquatic organisms are strongly dependent on chemical speciation [7]. The free ion activity model (FIAM) and the biotic ligand model (BLM) can be used to predict this bioavailability. Their

Chapter 2

applicability has been demonstrated for algae in defined media exposed to Cd [74, 97, 98], although some exceptions have been found [64, 77, 78].

Periphyton is the most important primary producer in running waters and responsible for the uptake and retention of organic carbon and inorganic nutrients [47]. It is a complex, three dimensional community of heterotrophic and autotrophic organisms (algae, bacteria, fungi) and non-living components, which grows on various substrates in rivers, lakes and wetlands [47]. The community is embedded in a heterogeneous matrix, composed of different macromolecules such as polysaccharides, proteins, nucleic acids, (phospho)lipids and other polymeric compounds [50]. It has been shown in some laboratory [99-101] and field studies [5, 86] that periphyton is very effective in accumulating metals from water. Elevated Cd concentrations in water can have effects on biomass related parameters (chlorophyll *a*, dry weight) [60, 61] and on settlement, development and species composition of periphyton [62]. There is also evidence of Cd biomagnification among trophic levels [63]. Periphyton can accumulate metals by three different processes, namely adsorption to components of the extracellular matrix, adsorption to cell surface molecules and intracellular uptake [47]. Kinetic studies with planktonic algae cultures under laboratory conditions showed a rapid increase of Cd uptake and a slow release, when cells were transferred to Cd free medium [74, 77]. It is expected that accumulation of Cd in periphyton under freshwater conditions will differ from Cd uptake by planktonic algae in defined media, since periphyton includes different algal species, which are embedded in an organic matrix. Only few studies are available investigating the uptake and release of Cd in periphyton [92, 93, 101], which in particular do not give information about the intracellular Cd content in combination with exposure to natural freshwater. Uptake kinetics of Cd in periphyton are of interest in order to evaluate the effects of Cd concentration variations in natural waters.

The aim of the present work was to study the kinetics of Cd accumulation (total and intracellular) in periphyton under freshwater conditions exposed to environmentally relevant Cd concentrations in a short-term microcosm experiment. Periphyton was precolonised in artificial flow-through channels supplied with natural freshwater and then exposed to nominal Cd concentrations of 5 and 20 nM for 26.4 hours. Cadmium accumulation was related to either dissolved or labile Cd concentrations in water. The latter was measured *in situ* with the technique of diffusion gradients in thin-films (DGT) [40, 102], allowing the simultaneous measurement of average concentrations of free, inorganic metal complexes and some part of the organic metal complexes over the deployment time. Intracellular Cd uptake in periphyton during Cd exposure was modelled with a first-order one compartment model. The release of Cd from periphyton upon cessation of Cd exposure was also investigated. Uptake and release kinetics were compared with data from algae and periphyton experiments, and factors affecting the accumulation of Cd in periphyton are discussed.

2.3. Materials and methods

2.3.1. Channels setup for periphyton colonisation

Periphyton was colonised in three artificial flow-through channels (86 cm long, 10.4 cm wide and 10 cm high, polymethyl methacrylate (PMMA)) which had a premixing chamber for the incoming water [103]. Sixty four microscope glass slides were fixed pairwise and vertically (long edge) in four rows in grooves along each channel to avoid the coverage of periphyton with sediment. Water flow in the channels was 4 L/min with a water depth of approximately 3.1 cm and a water column of around 0.5 cm above the slides. The channels were supplied with natural freshwater from the nearby Chriesbach stream containing low dissolved Cd background concentrations and the organisms for the colonisation of periphyton. Illumination was provided by two PAR-lamps (Osram HQL (MBF-U) Deluxe 400 W) overhanging the channels. The photoperiod was 15 h darkness, 9 h light and the average light irradiance 329 ± 59 µmol photons/m^2 X sec.

2.3.2. Kinetic experiment

Prior to the experiment periphyton was colonized for three weeks, which resulted in sufficient biomass to measure metal content and a homogeneous coverage of slides with periphyton. Experiments were run in the colonisation channels, using the same water flow, light intensity and photoperiod conditions as during colonisation. Prior to the start of the kinetic experiment two solutions of 405 and 1620 nM Cd were prepared in 200 L polyethylene barrels (Semadeni). Cadmium nitrate from a concentrated standard solution (J. T. Baker) was added to Chriesbach stream water which was previously diluted 1:27 with nanopure water to avoid precipitation of cadmium carbonate. Solutions were equilibrated for 19 hours to allow full equilibration of metals with natural ligands. In order to obtain nominal Cd concentrations of 5 and 20 nM in the channels, Cd solutions were added with a peristaltic pump (MCP SW 5.01, Ismatec, dilution ~1: 67) to plastic tubes supplying the channels with natural water. To ensure further mixing of the Cd solutions with the natural water, a plate with three perforated vertical plates was placed in the premixing chamber of each channel. Periphyton was exposed in two channels for 26.4 h to nominal Cd concentrations of 5 and 20 nM in Chriesbach water to study the kinetics of uptake (exposure period). After Cd addition was halted, periphyton was exposed to the original water from stream Chriesbach containing low dissolved Cd background concentrations for 21.2 h to investigate the release of Cd (postexposure period). One channel without Cd addition was used as the control. Dark periods started 6.6 h and 3.8 h after the beginning of the exposure and postexposure period respectively. Channels were covered with black plastic during this time to avoid any light influence on periphyton. Periphyton slides and water samples for dissolved metal concentrations were taken four times during the first 71 min after the start of the exposure period to examine the initial uptake of Cd in periphyton, four times during the subsequent exposure period and once at the end of the postexposure period.

2.3.3. Clean trace metal handling

To avoid contamination of samples and equipment plastic gloves (Semadeni) were used for all procedures. All channel equipment, barrels, polypropylene vials, bottles and beakers, DGT devices and filters, syringes and filtration units were placed in 0.1 M HNO_3 for at least 24 h and then properly rinsed with nanopure water and when necessary sealed in clean plastic bags. Cellulose nitrate filters (0.45 µm, Sartorius) for metal content in periphyton were boiled twice in 0.1 M HNO_3, rinsed with nanopure water, dried twice afterwards at 50°C for 15 hours in an oven and then preweighed. Except for the sampling at the channels all handling was performed in a clean bench.

2.3.4. Water sampling for dissolved metal concentrations

Water samples (14 ml) for determination of dissolved metal concentrations were collected at the end of the channels and filtered (0.45 µm filters, Milipore) into polypropylene tubes using a plastic syringe (BD Plastipak, 50 ml). Before sample collection filters and syringes were thoroughly rinsed with water from the channel. Samples were acidified to 0.24 M with HNO_3 (65% suprapure, Merck) and kept at 4°C in the dark till analysis. Water samples from the Cd solutions were taken following the same procedure.

2.3.5. Preparation, sampling and processing of DGT devices

Labile metal concentrations in water were measured with DGT (Diffusion gradients in thin-films). The procedure for making DGT devices followed the recommendation by Zhang and Davison [102]. Non-restricted diffusive gel (pore size ≈ 5 nm) with a thickness of 0.8 mm and resin hydrogel with a thickness of 0.4 mm were covered with acid-cleaned 0.45 µm cellulose nitrate filters and enclosed on a piston with a cap (high-density polypropylene). At the beginning of the exposure period three DGT devices were placed at the end of each channel, floating on the water surface with the resin pointing down. At the end of the exposure period they were removed and replaced with three new DGT devices for the postexposure period. For measurement of labile metal concentrations, the resin gel layer was removed and placed for 24 h in a 14 ml polypropylene vial containing 2 ml of 1.66 M HNO_3 (65% suprapure, Merck) and then diluted sevenfold. To calculate DGT-labile metal concentrations in water, average water temperatures during the time of deployment were used to obtain diffusion coefficients for free metal ions. The calculations for labile metal concentrations are described by Zhang and Davison [39] and the diffusion coefficients for free metal ions were measured by Hao Zhang, DGT Research Ltd., Lancaster (personal communication).

Free Cd concentrations in water were estimated using the speciation program vMINTEQ (http://www.lwr.kth.se/english/OurSoftware/vminteq/) [29]. Because dissolved organic carbon was not considered in the program, DGT-labile Cd concentrations were assumed to correspond only to inorganic complexes and were used as total dissolved Cd concentrations as input parameters for the calculation together with average concentrations of major cations, anions and alkalinity, pH and temperature for the exposure (four samples) and postexposure period (two samples).

2.3.6. Periphyton sampling and processing

For metal analysis, six periphyton slides were sampled randomly from each channel and stored in a plastic box till further processing. Periphyton was scratched from the slides with a microscope slide into a plastic beaker containing 100 ml of filtered experimental water. To obtain homogeneous periphyton suspensions, stirred solutions were first mixed using a 5 ml pipette and then sonicated in an ultrasonic bath for 30 s. Sediment and particles were then allowed to settle and the supernatant containing periphyton organisms and the matrix was transferred into a plastic beaker and filled with filtered experimental water to 150 ml. From each suspension three measurement replicates of 20 ml were filtered through cellulose nitrate filters to obtain dry weight and total metal contents. It was assumed that the soluble part of the matrix was removed from the organisms by this treatment procedure. The rest of the suspension was treated for 10 min with 1 ml of 0.26 M EDTA (4 mM final concentration) to remove metals adsorbed to the cell walls. The remaining metal content is considered to be intracellular (EDTA-non-exchangeable) [86]. Three measurement replicates of 20 ml were then filtered to obtain intracellular metal contents. Filters were dried twice for 15 h at 50 °C in an oven and weighed. Afterwards they were digested in Teflon beakers with 4 ml of nitric acid (65% suprapure, Merck) and 1 ml of hydrogen peroxide (30% suprapure, Merck) for 24 min in a high-performance microwave digestion unit (MLS 1200 Mega, Sarasin). Solutions of digested filters were transferred into 25 ml Erlenmeyer flasks and diluted with nanopure water. The digestion procedure was also used for plankton reference material (CRM 414, Institute for reference materials and measurements, European Commission, Belgium) and blank filters. The measured metal content in periphyton was related to the measured dry weight.

Chlorophyll *a* was extracted from 5 ml of the suspension with ethanol and concentrations were measured using the HPLC method described by Murray et al. [104]. For determination of species composition 5 ml from suspensions were fixed with 4% formaldehyde. It was determined semi-quantitatively upon visual observation, using an inverted phase contrast microscope with a magnification of 640. A three-graded scale was used for the abundance (3: dominant, 2: moderately abundant, 1: rare).

2.3.7. Metal analysis

Metal concentrations were measured with high resolution inductively coupled plasma mass spectrometry (HR-ICP-MS) (Element 2, Thermo Finnigan). The accuracy of the ICP-MS measurements was checked using SLRS-4 (National Research Council Canada, errors: Cd < 8%, Zn < 9%, Mn < 6%, Cu < 10%), TM-28.2 (National Research Council Canada, errors: Cd < 6%, Zn < 9%, Mn < 9%, Cu < 9%) reference water and plankton reference material (errors: Cd < 10%, Zn < 7%, Mn < 7%, Cu < 10%).

2.3.8. Water chemistry

Water samples for DOC, alkalinity, major cation and anion analysis were taken four times during the exposure and once at the end of the postexposure period from the tank supplying the channels with water. Concentrations of major cations were determined by ICP-OES and anion concentrations were measured by ion chromatography (Metrohm). Alkalinity measurements were performed by titration (with HCl 0.1M until pH 4.5), and DOC was obtained by combustion. Temperature and pH were measured at the times of periphyton sampling.

2.3.9. Modelling of intracellular cadmium uptake and release in periphyton

Initial intracellular uptake rates (r_i) in periphyton were modelled linearly during the first 71 min of Cd exposure at both exposure concentrations. Kinetics in this model are dictated only by uptake. To model the intracellular Cd uptake during the whole exposure period a nonlinear one compartment model was used, which assumes that uptake and release follow first order kinetics:

$$\{Cd_{periphyton}\} = \frac{k_1}{k_2} \times [Cd_{water}] \times (1 - e^{-k_2 t}) \quad (1)$$

where k_1 [L/g dw X min] and k_2 [1/min] are Cd uptake and clearance rate constants for periphyton, $Cd_{periphyton}$ [nmol Cd/g dw] is the intracellular concentration of Cd in periphyton, Cd_{water} [nmol Cd/L] the Cd concentration in water and t [min] is the time of Cd exposure. The modelling was performed with the program GraphPad Prism 4.03. Computed parameters were intracellular Cd concentrations in periphyton at the end of the exposure period ($\{Cd_{max,\ predicted}\}$), represented by the term $\frac{k_1}{k_2} \times [Cd_{water}]$ in equation 1 and clearance rate constants k_2. Uptake rate constants k_1 were calculated from these parameters using dissolved and labile Cd concentrations in water ($[Cd_{water}]$).

Average dissolved and labile Cd concentrations in water (Cd_{water}) during the exposure period were used together with the corresponding uptake rate constants k_1 to calculate average uptake rates ($\bar{r}_1$):

$$\bar{r}_{1, dissolved/labile} = k_1 \times [Cd_{water}] \quad (2)$$

Intracellular Cd concentrations in periphyton at the end of the postexposure period $\{Cd_{end, experimental}\}$ were compared with intracellular concentrations predicted with a first-order rate release model using clearance rate constants k_2 obtained from the modelled uptake:

$$\{Cd_{end, predicted}\} = \{Cd_0\} \times (e^{-k_2 t}) \quad (3)$$

where $Cd_{end, predicted}$ [nmol Cd/g dw] is the intracellular Cd concentration in periphyton at the end of the postexposure period, Cd_0 [nmol Cd/g dw] the intracellular Cd concentration in periphyton at the end of the exposure to Cd, k_2 the calculated clearance rate constant and t the time of the postexposure period.

Bioconcentration factors (BCF) at steady state [L/kg] were calculated for the two exposures by:

$$BCF(I)_{dissolved/labile} = k_1 / k_2 \quad (4)$$

where k_1 and k_2 are the uptake and clearance rate constants and k_1 being calculated with dissolved or labile Cd concentrations in water. Bioconcentration factors were also calculated for Cd exposed and control periphyton by:

$$BCF(II)_{dissolved/labile} = \{Cd_{intracellular}\} / \{Cd_{water}\} \quad (5)$$

where $Cd_{intracellular}$ {nmol Cd/kg dw} is the intracellular Cd content in periphyton at the end of the exposure period and Cd_{water} [nmol Cd/L] are either the dissolved Cd concentrations at the end of the exposure period or labile Cd concentrations during the exposure period.

2.4. Results

2.4.1. Dissolved and DGT-labile metal concentrations in water

Dissolved, DGT-labile and free Cd concentrations in water as well as percentages of DGT-labile and free Cd during the exposure and postexposure period for all channels are summarized in Table 2.1. Average background concentrations of dissolved Cd in water of the control channel were low (0.12 ± 0.02 nM) during the exposure period. Two rain events during the second light and dark period elevated average concentrations to 0.2 ± 0.06 nM during the postexposure period. The percentages of DGT-labile (11-13%) and free Cd (8-9%) in the control channel were low during the whole experiment. In both Cd exposure channels, dissolved Cd concentrations remained constant during the exposure period (5 nM: 5.8 ± 0.3, 20 nM: 25 ± 3). Compared to the control high and similar proportions of DGT-labile (5 nM: 60%, 20 nM: 69%) and free Cd (5 nM: 43%, 20 nM: 49%) were determined in the Cd exposure channels. Shortly after the end of the exposure period, dissolved Cd concentrations were comparable to the control (20 nM exposure: 0.26 nM, 5 nM exposure: 0.18 nM, control: 0.16 nM). During the postexposure period average concentrations of dissolved and DGT-labile Cd in the Cd exposure channels were as low as in the control.

Dissolved Zn, Cu and Mn concentrations in water were similar in all channels during the exposure (Zn: ~240 nM, Cu: ~32 nM, Mn: ~93 nM) and postexposure period (Zn: ~463 nM, Cu: ~53 nM, Mn: ~125 nM) (Table 2.2). The two rain events also elevated dissolved concentrations of these metals during the postexposure period. Percentages of DGT-labile Zn in water were around 36% during the whole experiment in all channels. Percentages of DGT-labile Cu increased with increasing Cd concentrations in water during the exposure period, namely from 24% in the control, to 28% in the 5 nM and 33% in the 20 nM exposure. During the postexposure period percentages of DGT-labile Cu were around 40% in all channels. Although labile Mn concentrations were not measured, data obtained in other studies with Chriesbach water (unpublished data) showed that most of total dissolved Mn is labile.

Chapter 2

Dissolved organic carbon was 2.68 ± 0.1 mg/L, calcium 2.89 ± 0.06 mM and alkalinity 6.28 ± 0.15 mM during the exposure period (means of four samples). The rain events caused an increase of DOC to 3.07 mg/L and a decrease of calcium to 2.58 mM and alkalinity to 5.65 mM at the end of the postexposure period. Temperature and pH were similar and constant in all channels over the whole experiment.

Exposure period

Exposure	Diss. Cd	Labile Cd	% labile Cd	Free Cd	% free Cd	pH	Temp.
Control	0.12 ± 0.02	0.013 ± 0.015	11	0.009	8	7.93 ± 0.03	9.7 ± 0.4
5 nM Cd	5.8 ± 0.3	3.5 ± 0.6	60	2.5	43	7.97 ± 0.06	9.8 ± 0.4
20 nM Cd	25 ± 3	17.2 ± 0.4	69	12.2	49	7.94 ± 0.06	9.9 ± 0.4

Postexposure period

Exposure	Diss. Cd	Labile Cd	% labile Cd	Free Cd	% free Cd	pH	Temp.
Control	0.20 ± 0.06	0.025 ± 0.006	13	0.018	9	7.92 ± 0.01	9.5 ± 0.1
5 nM Cd	0.21 ± 0.04	0.030 ± 0.024	14	0.021	10	7.90 ± 0.06	9.6 ± 0.1
20 nM Cd	0.25 ± 0.01	0.021 ± 0.000	8	0.015	6	7.92 ± 0.04	9.5 ± 0.1

Table 2.1: Dissolved [nM], DGT (diffusion gradients in thin films)-labile [nM] and free Cd [nM] concentrations in water, percentages of DGT-labile and free Cd as well as pH and temperature [°C] during the exposure and postexposure period in all channels. Values represent means ± standard deviation. Means represent eight samples respectively measurements for dissolved Cd concentrations, pH and temperature during the exposure period, two during the postexposure period and three DGT replicates for each period.

Exposure period

Exposure	Diss. Zn	Labile Zn	% labile Zn	Diss. Cu	Labile Cu	% labile Cu	Diss. Mn
Control	252 ± 65	94 ± 11	37	32 ± 4	7.7 ± 1.4	24	94 ± 6
5 nM Cd	214 ± 41	78 ± 19	36	31 ± 4	8.7 ± 1.9	28	91 ± 9
20 nM Cd	256 ± 41	92 ± 7	36	31 ± 3	10.2 ± 1.9	33	93 ± 7

Postexposure period

Exposure	Diss. Zn	Labile Zn	% labile Zn	Diss. Cu	Labile Cu	% labile Cu	Diss. Mn
Control	463 ± 102	144 ± 19	31	53 ± 3	22.8 ± 5.3	43	126 ± 33
5 nM Cd	458 ± 116	173 ± 21	38	53 ± 5	22.4 ± 3.0	43	121 ± 22
20 nM Cd	468 ± 98	164 ± 41	35	54 ± 2	18.7 ± 1.8	35	127 ± 25

Table 2.2: Dissolved [nM] and DGT (diffusion gradients in thin films)-labile [nM] concentrations of Zn and Cu and dissolved [nM] concentrations of Mn in water as well as percentages of DGT-labile concentrations during the exposure and postexposure period in all channels. Representation of values and means are as described in Table 2.1.

2.4.2. Periphyton characterization

Chlorophyll *a* content of periphyton expressed as [mg Chl a/g dw] was 4.9 ± 1.4 in the control, 4.3 ± 1 in the 5 nM and 5.5 ± 1.4 in the 20 nM Cd exposure. Semi-quantitative microscopical analysis showed that periphyton was mostly composed of diatoms. The dominant species was *Nitzschia palea*, moderately abundant were *Achnanthes minutissima, Achnanthes lanceolata, Gomphonema parvulum* and rare were *Melosira varians, Cymbella affinis, Rhoicosphenia abbreviata, Nitzschia linearis, Nitzschia acicularis, Amphora ovalis, Cocconeis placentuala, Navicula rhynchocephala* and *Fragilaria capucina*. Chlorophyll *a* content as well as species composition and abundance were similar in all channels and did not change after exposure to Cd.

2.4.3. Kinetics of cadmium uptake and release in periphyton

Average total and intracellular Cd concentrations in control periphyton were low during the whole exposure period (Total: 0.97 ± 0.23 nmol Cd/g dw, intracellular: 0.65 ± 0.25 nmol Cd/g dw). At the end of the postexposure period total concentrations increased slightly to 1.9 ± 0.1 and intracellular to 1.2 ± 0.1 nmol Cd/g dw concomitantly to the elevation of dissolved Cd concentrations in water (Figure 2.1 a). Intracellular Cd concentrations increased steadily during the whole time of Cd exposure in the 5 and 20 nM treatments. In contrast, total Cd concentrations in both treatments increased steadily to the end of the first dark period, but decreased afterwards to the end of Cd exposure (Figures 2.1 b and c). At the end of the exposure period Cd content in periphyton was 21.9 ± 0.2 nmol Cd g/dw in the 5 nM, 85.7 ± 0.7 nmol Cd/g dw in the 20 nM exposure and intracellular was 13 ± 0.8 nmol Cd/g dw and 55.5 ± 2.7 nmol Cd/g dw respectively. The intracellular Cd content of periphyton at the end of the exposure period was 15 times higher in 5 nM and 64 times higher in the 20 nM treatment when compared to the control. Total and intracellular Cd concentrations in periphyton increased rapidly during the first 71 min (initial uptake). The subsequent total and intracellular content increase for the rest of the exposure period was slower than the initial uptake. Total and intracellular Cd contents increased also during the dark period in both treatments during the exposure period. Total and intracellular Zn, Cu and Mn contents in periphyton during the exposure period were similar in all channels (data not shown). After Cd addition was halted, total and intracellular Cd concentrations in periphyton decreased slowly, whereas total contents showed a faster decrease than intracellular contents.

The rapid increases of intracellular Cd concentrations in periphyton during the first 71 min of Cd exposure were fitted to a linear uptake (Initial uptake, Figures 2.1 b' and c'). Initial intracellular uptake rates were 0.05 nmol Cd/g dw X min in the 5 nM and 0.18 nmol Cd/g dw X min in the 20 nM exposure.

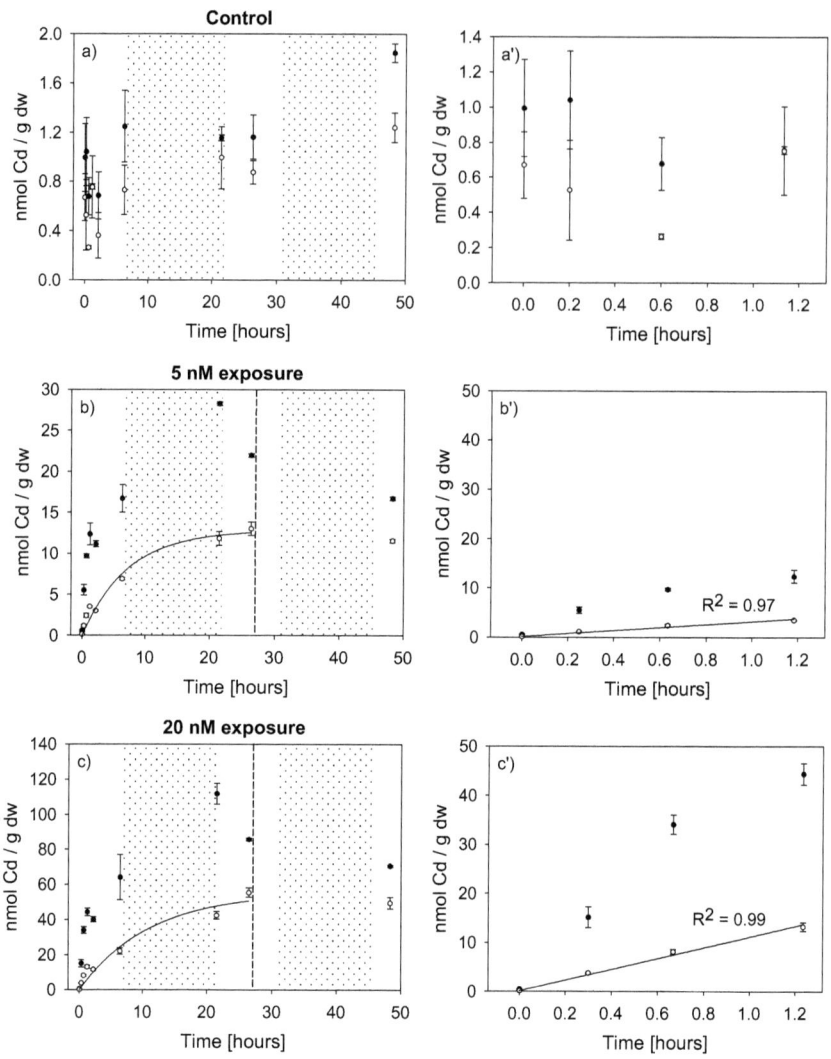

Figure 2.1: Total and intracellular Cd concentrations in periphyton during the exposure and postexposure period in all channels (a-c). The solid curves are the results from the non-linear model for intracellular Cd uptake during the exposure period. Initial Cd uptake in periphyton during the first 71 min of Cd exposure (a'-c'). The solid lines represent the modelled linear increase. Black dots represent total, white dots intracellular Cd concentrations in periphyton, dashed lines the end of Cd exposure and dotted rectangles the dark periods. Data points for Cd content in periphyton represent three replicates (means ± standard deviations).

Percentages of adsorbed Cd were similar at both exposure concentrations during the whole experiment and ranged from 58 to 79% during the exposure period except for the last point with a percentage of around 38%. At the end of the postexposure period the percentage of adsorbed Cd was around 30%, similar to the control.

2.4.4. Linear and non-linear modelling of cadmium accumulation in periphyton

Ratios of initial intracellular uptake rates (r_i) between the 20 and 5 nM exposure were compared with ratios of dissolved and labile Cd concentrations in water. Ratios were 3.6 for initial uptake rates, 3.8 for dissolved and 4.9 for labile Cd concentrations (Table 2.3a).

The non-linear model fitted the data well (R^2=0.96 (5 nM), R^2=0.95 (20 nM)). All modelled and calculated parameters are summarized in Table 2.3b. Values of uptake rate constants $k_{1,\ dissolved/labile}$ and clearance rate constants k_2 were similar at both exposure concentrations and values for k_1 were consistent for both calculations used. Ratios of averaged intracellular uptake rates ($\overline{r}_{1,dissolved}$, $\overline{r}_{1,labile}$) between the two treatments were 2.8 for both calculations. Averaged intracellular uptake rates were smaller than initial intracellular uptake rates (r_i). Values of bioconcentration factors were in the similar order of magnitude at both exposure concentrations and were consistent for all calculations used. Predicted {$Cd_{max,\ predicted}$} and experimental {$Cd_{max,\ experimental}$} intracellular Cd concentrations in periphyton at the end of the exposure period show equal values at both exposure concentrations. Comparisons of modelled intracellular Cd concentrations in periphyton at the end of the postexposure period {$Cd_{end,\ predicted}$} with the experimental values {$Cd_{end,\ experimental}$} show that Cd is released much slower at both concentrations than predicted by the model.

	Parameter	5 nM Cd	20 nM Cd	Ratio 20 : 5
a)	Dissolved Cd [nmol/L]	6.1 ± 0.1	23.4 ± 0.5	3.8
	Labile Cd [nmol/L]	3.5 ± 0.6	17.2 ± 0.4	4.9
	r_i: [nmol Cd/g dw X min]	0.05	0.18	3.6
b)	$k_{1,\,dissolved}$: [L/g dw X min]	0.0051	0.0033	0.65
	$\bar{r}_{1,dissolved}$: [nmol Cd/g dw X min]	0.03	0.08	2.8
	$k_{1,\,labile}$: [L/g dw X min]	0.0084	0.0048	0.57
	$\bar{r}_{1,labile}$: [nmol Cd/g dw X min]	0.03	0.08	2.8
	k_2: [1/min]	0.0023	0.0015	0.65
	$BCF(I)_{dissolved}$ [L/kg]	2.2×10^3	2.2×10^3	1
	$BCF(I)_{labile}$ [L/kg]	3.6×10^3	3.2×10^3	0.88
	$BCF(II)_{dissolved}$ [(nmol Cd/kg dw)/(nmol $Cd_{dissolved}$/L)]	2.5×10^3	1.9×10^3	0.76
	$BCF(II)_{labile}$ [(nmol Cd/kg dw)/(nmol Cd_{labile}/L)]	3.7×10^3	3.2×10^3	0.86
	R^2	0.96	0.95	-
	$Cd_{max,\,predicted}$ [nmol Cd/g dw]	13	56	4.4
	$Cd_{max,\,experimental}$ [nmol Cd/g dw]	13	56	4.3
	$Cd_{end,\,predicted}$ [nmol Cd/g dw]	0.64	7.9	12.4
	$Cd_{end,\,experimental}$ [nmol Cd/g dw]	11.6	49.6	4.3
	Modelled half life of Cd in periphyton [min]	301	466	1.6

Table 2.3: a) Dissolved and DGT (diffusion gradients in thin films)-labile Cd concentrations in water as well as initial uptake rates (r_i) in periphyton during the first 71 min of Cd exposure. Means ± standard deviations represent 3 samples for dissolved and three DGT replicates for labile Cd concentrations. Values for DGT-labile Cd concentrations are average concentrations from the exposure period. b) Results of experimental and modelled parameters for the uptake and release of Cd in periphyton. Parameters represent uptake rate constants (k_1), averaged uptake rates ($\bar{r}_1$), clearance rate constants (k_2), bioconcentration factors calculated from uptake and clearance rate constants (BCF(I)) and bioconcentration factors calculated from Cd concentrations in periphyton and water (BCF(II)). Some of the parameters were calculated with dissolved and labile Cd concentrations in water (subscript). Further parameters were correlation coefficient from the model (R^2), modelled and experimental Cd concentrations in periphyton at the end of the exposure period ($Cd_{max,\,predicted}$, $Cd_{max,\,experimental}$) and at the end of the postexposure period ($Cd_{end,\,predicted}$, $Cd_{end,\,experimental}$). Ratio 20:5 is the ratio of parameters between the 20 and 5 nM Cd exposure.

2.5. Discussion

2.5.1. Metal speciation in channels

Background concentrations of dissolved Cd in Chriesbach water were low as reported for other unpolluted sites [11, 12, 26]. The low percentages (11-13%) of DGT-labile Cd concentrations measured in the control channel indicate that most Cd was bound to non-labile organic ligands or colloids. Higher percentages of DGT-labile Cd (70%) were measured in a freshwater stream with similar dissolved Cd but lower dissolved organic carbon concentrations than in our study [11, 12, 26]. However, percentages of DGT-labile Cd were high in the Cd exposure channels, most probably because Cd solutions added to the channels contained an excess of Cd compared to organic ligands and other metals. The short contact time (~40 s) of Cd solutions with organic ligands of Chriesbach water in the channels was too short to achieve a new, complete chemical equilibrium, resulting in higher percentage of labile Cd concentrations. The increase of labile Cu concentrations with increasing Cd concentrations might be due to competition of Cd with Cu for binding sites of non-labile ligands. This effect was not observed with Zn, maybe due to the high concentrations of DGT-labile Zn. These results show that elevated Cd concentrations can change the speciation of other metals.

2.5.2. Cadmium accumulation and kinetics in periphyton

Two uptake phases for the intracellular Cd uptake, a fast initial one and a slower subsequent one reaching steady state at the end of the exposure period, were observed in periphyton in both Cd exposure channels. The time course of adsorbed Cd shows a continuous binding of Cd to cellular binding sites, which become saturated towards the end of the exposure period. The fast initial linear Cd uptake during the first 71 min of Cd exposure is assumed to be a consequence of uptake only. The subsequent slower uptake phase suggests that some Cd is released and a steady-state is reached. These results may be compared with Cd uptake kinetics in algae cultures under defined chemical conditions, in which uptake is only controlled by transport over the

Chapter 2

cell wall and membrane. The comparison shows that both the saturation of binding sites (adsorbed Cd) [64] and uptake is faster than in periphyton. In the green alga *Selenastrum capricornutum* Printz the initial uptake phase was 9 min and steady state was reached within 30 min at comparable free Cd concentrations [77]. Other than in free floating algae, the matrix and the three dimensional structure of the periphyton community may slow down the diffusion of Cd to cellular binding sites of algae.

Few data on Cd accumulation kinetics in periphyton are available for comparison. Cadmium concentrations of the whole periphyton biofilm in a lake exposed to 0.8 nM Cd increased nonlinearly, approaching steady state after 11 days of exposure [92]. Hill et al. [101] demonstrated a linear increase of Cd in the whole periphyton biofilm exposed to 8.9 nM Cd in flow-through channels supplied with natural freshwater over 48 hours. Comparisons among such studies are difficult, since various factors such as species composition and concentrations respectively speciation of Cd as well as of other metals influence the uptake of metals in periphyton in natural freshwaters. Moreover, the diffusion of Cd from water to the algae of periphyton is influenced by the thickness and composition of the matrix, sediments and particles incorporated in the matrix and water current [101].

The intracellular uptake of Cd in periphyton could be modelled with a simple, nonlinear one-compartment model. However, the used model does not provide a distinction between the diffusion kinetics (diffusive layer, matrix, cell wall) and the transport kinetics through the cell membrane. Uptake rate constants k_1 depend on diffusion of Cd through the diffusive boundary layer, the matrix and the transport across the plasma membrane, clearance rate constants k_2 from the binding to intracellular ligands and the transport across the plasma membrane. The consistent values of both constants show that they are independent of Cd concentrations in water. Stephenson and Turner [92] used a similar model for the accumulation of Cd in the whole periphyton biofilm exposed in lake water to 0.8 nM Cd and also obtained a good fit. They found a similar uptake rate constant of 0.026 L/g C min (calculated from [92]), when our results were referred to a carbon content of 16%, as obtained in previous experiments.

Initial uptake rates r_i increased with increasing Cd concentrations in water. Ratios of initial uptake rates were closer to ratios of dissolved than to those of labile Cd concentrations in water. Due to only two Cd exposure concentrations used and high labile Cd concentrations being close to dissolved concentrations, no clear relationship could be obtained between Cd accumulation in periphyton and Cd speciation in water. Averaged uptake rates ($\overline{r}_{I,dissolved}$, $\overline{r}_{I,labile}$) within the same exposure were lower than initial uptake rates, resulting from the influence of Cd release during the exposure. Stephenson and Turner [92] found averaged uptake rates of 0.021 nmol Cd/g C min at Cd concentrations of 0.8 nM. In our study they were 0.188 and 0.500 nmol Cd/g C min in the 5 nM and 20 nM Cd exposures, respectively (a carbon content of 16% was assumed as obtained in previous experiments). These uptake rates are thus similar, if related to the Cd concentrations in water.

Values of bioconcentration factors (BCF) for the two Cd exposure concentrations were in the similar order of magnitude and consistent for all calculations used. Bioconcentration factors in periphyton found by Morin et al. [61] were higher than in our study, probably due to exposure in media without competing organic ligands and excess of other metals. Bioconcentration factors calculated from Cd concentrations in periphyton and water (Table 2.3), were higher in the control channel (BCF(II)$_{dissolved}$: 6.6 X 10^3 L/kg, BCF(II)$_{labile}$: 6.7 X 10^4 L/kg) than in the Cd exposure channels, showing a decrease in bioconcentration factors with increasing Cd concentrations. This slight decrease might be due to an active regulation of internal Cd concentrations, i.e. an active release of Cd at the two exposure concentrations. Decreases of BCFs with increasing Cd concentrations in water were also found for periphyton exposed to Cd concentrations of 89 and 890 nM [61] and also reported for different algal species [105].

2.5.3. Competition of other metals for cadmium uptake

Despite the excess of total dissolved and labile Zn, Mn and Cu concentrations in water, total and intracellular Cd concentrations in periphyton increased during the exposure to Cd in the 5 and 20 nM treatments and in the control during the two rain events. Competitions of these metals for Cd uptake have been demonstrated with algae cultures in defined media. Cadmium uptake was inhibited by Zn and Mn in several marine diatoms [97, 106] and in *Scenedesmus vacuolatus* [107] and by Cu and Zn in *Chlamydomonas reinhardtii* [74]. Different uptake mechanisms in algae have been identified. In *Scenedesmus vacuolatus* for example binding constants of Cd and Zn to transport sites had similar values [107] and the Zn transport system was assumed to be the main uptake path for Cd. Cadmium uptake in algae can also occur via the Mn transport system [97, 106] or a divalent metal transporter [108] acting in the transport of Mn, Fe and Cu. The efficient Cd uptake in our study in the presence of an excess of competing metals indicates that Cd has a high affinity to transport sites of the plasma membrane. A faster Cd accumulation and higher Cd contents in periphyton at steady-state would be expected in the absence of those metals [97, 106].

2.5.4. Cadmium release

Total and intracellular Cd concentrations in periphyton decreased slowly after Cd addition was halted. Obtained clearance rate constants from the modelled uptake were higher when compared to published values for periphyton, being in the range of 2.03×10^{-4}/min [92] to 5.42×10^{-5}/min [93]. Predicted intracellular Cd concentrations in periphyton at the end of the postexposure period using a first-order decrease model and calculated clearance rate constants k_2 (Table 2.3b) were lower than measured, showing an overestimation of k_2 by the model. Such differences at the end of a Cd release experiment were also observed by Stephenson and Turner [92] for lake periphyton. The slow decrease and the overestimated clearance rate constants may be due to binding of Cd to intracellular ligands, for instance phytochelatins. Phytochelatins are cysteine-rich metal binding polypeptides which are strongly induced in algae by Cd even at low concentrations [89, 90], with synthesis starting within minutes to hours after exposure [109]. Phytochelatins are only one regulation mechanism within the complex network of metal homeostasis, which shows interspecific differences and may thus result in different residence times of intracellular Cd concentrations in periphyton.

As for the Cd uptake, differences were observed when the release from periphyton is compared with the release from planktonic algae [74, 77]. The slower release in periphyton might be due to a retention of released Cd by the matrix and a subsequent uptake by adjacent cells [93].

2.5.5. Environmental relevance

In terms of environmental relevance, our data show that periphyton accumulates Cd very rapidly at low Cd concentrations and that most Cd is retained in the cells when Cd exposure is halted. Although Cd content in periphyton can be reduced by release, growth dilution and cell desorption [93, 110], successive exposures to elevated metal concentrations in water may maintain high concentrations of Cd and other metals in periphyton. Due to its high toxicity and persistence Cd will affect directly periphyton and associated consumers. Cadmium released back to the water from periphyton may also affect other aquatic organisms.

Chapter 3

Accumulation of cadmium in periphyton under various freshwater speciation conditions

This chapter is published in *Environmental Science and Technology, 2009, 43: 7291–7296*

3.1. Abstract

The goal of the present study was to examine the relationship between Cd speciation and accumulation in periphyton at environmentally relevant Cd concentrations under natural freshwater conditions. Both equilibrium and non-equilibrium based models for the prediction of bioavailability of metals have been developed from experiments with algae in defined media. However, validation studies with multispecies communities like periphyton in natural freshwaters are scarce. Periphyton was exposed in artificial recirculating channels containing natural freshwater to two Cd concentrations (20 and 40 nM), for which speciation was changed by the addition of a synthetic organic ligand (NTA). Labile metal concentrations were measured with the technique of diffusion gradient in thin-films (DGT) and major Cd species were estimated by modelling. Total and intracellular Cd content in periphyton increased within both Cd exposure concentrations and could be related to an increase in DGT-labile Cd, which was caused by the competition of NTA with DGT-non labile ligands. Bioaccumulation was thus not controlled by the free Cd concentrations, as predicted by the equilibrium based models, but by the diffusion of labile Cd-NTA complexes, which constituted a large fraction of DGT-labile Cd.

3.2. Introduction

Bioavailability and hence uptake of essential and non-essential trace metals to aquatic organisms depend on the chemical speciation of the metals as well as on their concentrations [7, 14]. Models have been developed to relate biouptake to metal speciation in water. The free ion activity model (FIAM) predicts that uptake is controlled by the free metal ion concentration in solution [70]. The applicability of the FIAM has been demonstrated for algae in defined media [74, 97, 98], but exceptions have also been found [64, 77, 78]. The biotic ligand model (BLM) [64, 81], which is derived from the FIAM, takes other cations and natural ligands into account. Cations (metals, H^+, Ca^{2+}) can compete with metals for uptake sites of algae (biotic ligand) and natural ligands can alter the speciation of both cations and metals.

Both are equilibrium-based models assuming that 1) equilibrium is attained between the metal species in the bulk solution and that adsorbed to the biotic ligand and 2) the diffusion flux of metal species from the bulk solution to the plasma membrane is faster than the uptake flux. But both models do not consider the dynamic aspects of metal complexes, i.e. their mobility and lability. Non-equilibrium based models include diffusion and internalization fluxes. If diffusion of free metal ions from the bulk solution to the plasma membrane is rate limiting then metal uptake can be controlled by the labile metal species. This has been shown theoretically [23, 82], with algae in defined media in the case of Ag [85] and with periphyton in natural freshwaters at very low free metal ion concentrations [86].

The aim of this study was to investigate which cadmium species controls accumulation in periphyton under freshwater conditions. Modification of Cd speciation was achieved by adding different concentrations of an artificial organic ligand (nitrilotriacetic acid, NTA) to natural freshwater containing two environmental relevant Cd concentrations. It was expected that NTA would increase labile and decrease free Cd concentrations. NTA forms complexes with Cd which are fully labile and can be measured with the technique of diffusion gradients in thin films (DGT) [111]. Free Cd concentrations and other major Cd species were estimated with a speciation program, based on equilibrium with natural organic matter (NOM) and NTA. Experiments were carried out in artificial recirculating channels. Total and intracellular Cd concentrations in periphyton were related to total dissolved, DGT-labile and free Cd concentrations in water. The results were compared to the predictions of the equilibrium based and non-equilibrium based models.

3.3. Materials and methods

3.3.1. Periphyton colonization

Prior to the experiment, periphyton was colonised on glass slides in two artificial channels [103] for ten weeks and four DGT devices were deployed for 5.8 days. The channel setup is described in detail in a previous publication [13]. The channels were continuously supplied with freshwater from the nearby Chriesbach stream containing low dissolved Cd background concentrations and the organisms for the colonisation of periphyton. Water flow in the channels was 10 L min^{-1}. Illumination was provided by two PAR-lamps (Osram HQL (MBF-U) Deluxe 400 W) overhanging the channels. The photoperiod was 15 h darkness, 9 h light and the average light intensity 341 ± 48 µE m^{-2} sec^{-1}.

3.3.2. Experimental design

Periphyton slides were exposed in artificial channels with the water running in a closed circuit. Water flow was 6 L min^{-1}, light intensity 367 ± 24 µE m^{-2} sec^{-1} and illumination was continuous. Prior to the experiment three exposure solutions of 20 nM Cd (NTA: 0, 2×10^{-6}, 8×10^{-6} M) and three of 40 nM Cd (NTA: 0, 2×10^{-6}, 8×10^{-6} M) were prepared in 20 L canisters (PE-HD), by adding cadmium nitrate from a concentrated standard solution (J. T. Baker) to Chriesbach stream water. Canisters were sealed and placed in a basin with a through-flow of Chriesbach water to maintain constant temperature and equilibrated for 16 hours to allow full equilibration of metals with natural ligands and NTA. For the experiment solutions were purged with an air mixture (0.5% CO_2, 21.5% O_2, 78% N_2) to maintain constant pH, first for 30 min alone and then for 60 min with the channels running without periphyton slides and DGT devices. Afterwards 12 periphyton slides were fixed vertically in each channel and 8 DGT devices were placed at the end of each channel, floating on the water surface with the resin pointing down. DGT devices and periphyton slides were removed before the start of the exposure experiment and used as the control.

3.3.3. Clean trace metal handling

To avoid contamination of samples and equipment, plastic gloves (Semadeni) were used for all procedures. Samples and the material for sampling were sealed in plastic bags. The experimental system was protected by plastic coverings. Components of the channel system, boxes for periphyton transport, polypropylene vials, bottles and beakers, DGT devices, holders and filters, syringes and filtration units were placed in 0.1 M HNO_3 for at least 24 h and then properly rinsed with nanopure water. Cellulose nitrate filters for metal content determination in periphyton were boiled twice in 0.1 M HNO_3, rinsed with nanopure water and dried twice at 50°C for 15 hours in an oven. Except for the sampling at the channels all handling was performed in a clean bench.

3.3.4. Sampling

Six periphyton slides for total and intracellular metal content as well as for chlorophyll-*a* content and species composition and 4 DGT devices for labile metal concentrations in water were sampled after 3 and 6 hours of exposure. Water samples (14 ml) for dissolved metal concentrations were taken every hour at the beginning of the channels and filtered (0.45 µm filters, Milipore) into polypropylene tubes using a plastic syringe (BD Plastipak, 50 ml). Filters and syringes were previously thoroughly rinsed with water from the channels. Samples were acidified to 0.24 M with HNO_3 (65% suprapure. Merck) and kept at 4°C in the dark till analysis. Temperature and pH were measured manually at the same times. Water samples for alkalinity, major cation and anion analysis were taken at the start of the experiment and after 3 and 6 hours of exposure in each channel. Natural organic matter (NOM) was characterized in the 20 nM Cd exposure without NTA after 3 hours. Further water samples for total organic carbon content (TOC) were taken from some of other channels.

Water for dissolved metal concentrations was also sampled in the control channels during the time of DGT deployment with temperature and pH being measured at the same times.

3.3.5. NOM analysis and modelling of metal speciation

Natural organic matter was analysed by size-exclusion chromatography in connection to on-line, high sensitivity organic carbon detection (LC-OCD) [32-34]. Fractions determined by this method are biopolymers (polysaccharides, proteins), humic substances (humic and fulvic acids), degradation products of humic substances (building blocks), low molecular weight humics (degradation products of humic substances), low molecular weight acids (mono- and dicarboxylic acids) and neutrals/amphiphilics (amino acids, ketones, aldehydes, alcohols). Characterisation of molecular weight and aromaticity of humic substances, allowed further differentiation between fulvic and humic acids. Concentrations of characterised humic substances were used together with dissolved concentrations of metals (Cd, Zn, Mn, Cu, Pb, Fe), alkalinity, cations (Na$^+$, K$^+$, Ca^{2+}, Mg^{2+}) and anions (Cl$^-$, SO$_4^{2-}$, NO$_3^{2-}$, o-PO$_4^{3-}$), as well as pH and temperature to calculate concentrations of free metal ions, inorganic metal complexes, NTA-metal complexes and fulvic acid-metal complexes using the program vMINTEQ [29] with the Stockholm humic acid model [31]. Calculations were performed with the assumption of Fe being present as dissolved Fe(III).

3.3.6. Preparation and processing of DGT devices

Labile metal concentrations in water were measured with DGT (Diffusion gradients in thin-films). DGT-devices were made following the procedure described by Zhang and Davison [102], as described in detail in Bradac et al. 2008 [13]. For measurement of labile metal concentrations, the resin gel layer was removed and placed for 24 h in a 14 mL polypropylene vial containing 2 mL of 1.66 M HNO$_3$ (65% suprapure, Merck) and then diluted sevenfold. To calculate DGT-labile metal concentrations in water, average water temperatures during the time of deployment were used to obtain diffusion coefficients for free metal ions. The calculations for labile metal concentrations followed the procedure by Zhang and Davison [39]. Percentages of DGT-labile metal concentrations are related to averaged dissolved metal concentrations over the time of DGT deployment.

Since DGT measures concentrations of dynamic species and the concentrations of modelled metal species are based on equilibrium conditions, values can not be compared directly. Labile metal species can be calculated from the modelling results (Me_{labile}) by including diffusion coefficients of metal complexes:

$$Me_{labile} = \Sigma\ (Me^{z+} + Me_{inorg} + 0.2 \times (Me_{FA}) + 0.7 \times (Me_{NTA}))$$

where M^{z+} is the free metal ion concentration, M_{inorg} are inorganic metal complexes, M_{FA} are metal-fulvic acid complexes and M_{NTA} are metal-NTA complexes. Factors of 0.2 and 0.7 are diffusion coefficients of the corresponding Cd complexes when compared to diffusion coefficients of free metal ions [111, 112].

3.3.7. Periphyton processing

To avoid dehydration and contamination of periphyton slides, they were stored in a plastic box till further processing. The processing of periphyton samples and the procedure to discriminate between total and intracellular (EDTA-non-exchangeable) metal content is described in detail in Bradac et al. 2008 [13]. The measured metal content in periphyton was related to the measured dry weight.

Chlorophyll-*a* was extracted from 5 mL of the periphyton suspensions with ethanol [113] and concentrations were measured using the HPLC method described by Murray et al. [104]. For determination of species composition, 5 mL of periphyton suspensions were fixed with 4% formaldehyde and examined using an inverted phase contrast microscope with a magnification of 640, to provide a semi-quantitative estimate.

3.3.8. Analytical methods

Metal concentrations in water, DGTs and periphyton were measured with high resolution inductively coupled plasma mass spectrometry (HR-ICP-MS) (Element 2, Thermo Finnigan). The accuracy of the ICP-MS measurements was checked using the reference waters SLRS-4 (National Research Council Canada, errors: Cd < 11%, Zn < 12 %, Mn < 8%, Cu < 11 %, Pb < 12 %, Fe < 8 %) and TM-28.2 (National

Research Council Canada, errors: Cd < 5 %, Zn < 11 %, Mn < 6 %, Cu < 7 %, Pb < 8 %, Fe < 5 %) and plankton reference material (CRM 414, Institute for reference materials and measurements, European Commission, Belgium, errors: Cd < 12 %, Zn < 15 %, Mn < 13 %, Cu < 13 %, Pb < 14 %).

Concentrations of major cations were determined by ICP-OES and anion concentrations were measured by ion chromatography (Metrohm). Alkalinity measurements were performed by titration (with HCl 0.1M until pH 4.5).

3.4. Results

3.4.1. Dissolved and DGT-labile metal concentrations

Dissolved metal concentrations in the control channel were 0.11 ± 0.05 nM Cd (DGT-labile 43%), 182 ± 119 nM Zn (DGT-labile 46%), 95 ± 13 nM Mn (DGT-labile 82%), 38 ± 19 nM Cu (DGT-labile 45%), 0.33 ± 0.22 nM Pb (DGT-labile 35%) and 134 ± 33 nM Fe (DGT-labile 49%). Dissolved Cd concentrations were similar and constant in the three exposure channels with 20 nM Cd (17.3 ± 0.22 nM) as well as in the three with 40 nM Cd (33.8 ± 0.6 nM) (Figure 3.1 a). Dissolved concentrations of other metals were similar and constant in all exposure channels, namely Zn (258 ± 16 nM), Mn (83 ± 2 nM), Cu (40 ± 2 nM) and Pb (1.4 ± 0.5 nM). Dissolved Fe concentrations increased in both Cd exposures with increasing NTA concentrations (87 ± 4 nM (no NTA), 107 ± 3 nM (NTA: 2×10^{-6} M), 136 ± 10 nM (NTA: 8×10^{-6} M)). Percentages of DGT-labile Cd changed in the presence of NTA, but the effect was different in the two Cd exposures (Figure 3.1 a). DGT-labile Cd was similar to the control channel (43%) in the 20 nM and 40 nM Cd exposure without NTA and the 40 nM Cd exposure with 2×10^{-6} M NTA (~50%). But DGT-labile Cd increased to around 62% in the 20 nM Cd exposures in the presence of NTA and to 82% in the 40 nM Cd exposure with the highest NTA concentration. Compared to the control channel percentages of DGT-labile metals were higher in the Cd exposure channels for Zn (~98%), Mn (~146%), Pb (~101%) and Fe (~104%) and similar for Cu (~51%) (Figures 3.1 b, c and d-f). Some DGT-labile concentrations of

Zn, Pb and Fe exceeded dissolved concentrations maybe due to short DGT deployment times and contamination.

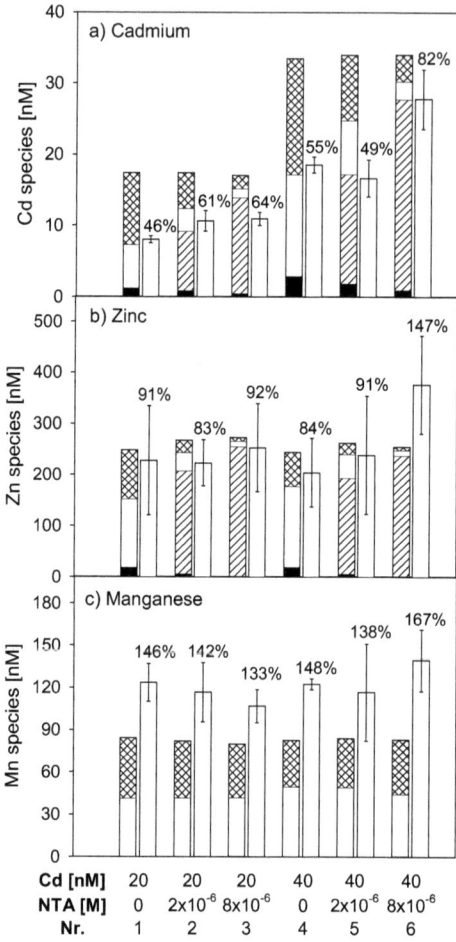

Figure 3.1 a-c: Speciation of cadmium, zinc and manganese. Dissolved metal concentrations in water of exposure channels after 6 hours (stacked bars) with concentrations of modelled metal species (Free metal ions (checkered texture), inorganic complexes (white texture), NTA-metal complexes (diagonal texture) and fulvic acid-metal complexes (black texture)). DGT-labile metal concentrations in water of exposure channels after 6 hours (plain white bars) as well as percentages of DGT-labile metals. Values of DGT measurements represent means ± standard deviations of 4 replicates.

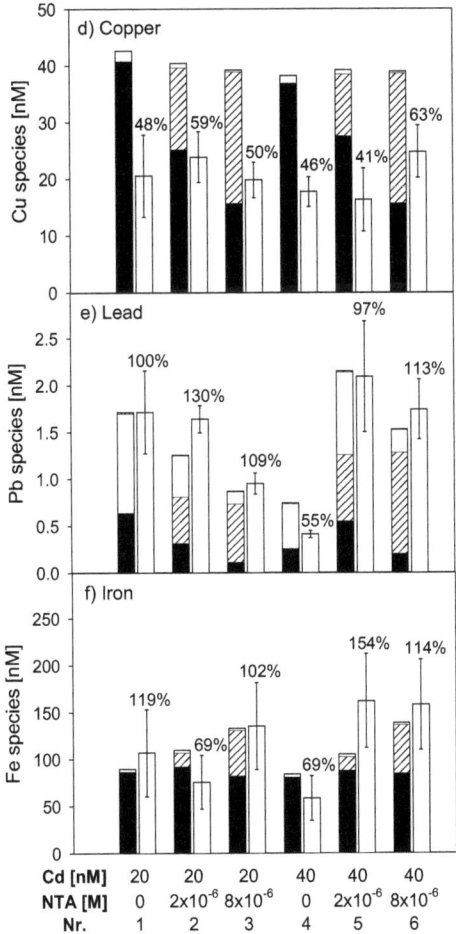

Figure 3.1 d-f: Speciation of copper, lead and iron. The different bars and textures represent metal speciation as described in Figure 3.1 a-c.

In the control channel alkalinity was 4.86×10^{-3} M, calcium 2.34×10^{-3} M, pH 8.02 ± 0.26 and temperature 11.2 ± 1.8 °C. These parameters were similar and constant in all Cd exposure channels, namely alkalinity was $5.97 \pm 0.01 \times 10^{-3}$ M, calcium $2.77 \pm 0.02 \times 10^{-3}$ M, pH 8.28 ± 0.08 and temperature 11.9 ± 0.2 °C.

3.4.2. NOM characterization

Total organic carbon (TOC) concentrations were constant at 2598 ± 21 µg C L^{-1} in the 20 nM Cd exposure channel with the highest NTA
concentration during 6 hours and comparable to the TOC content in the 40 nM Cd exposure with the same NTA concentration after 3 hours (2580 µg C L^{-1}). TOC content was also similar (2180 ± 26 µg C L^{-1}) at the two lower NTA concentrations after 3 hours.

Chromatographic fractions of hydrophilic dissolved organic carbon (DOC) determined in the 20 nM Cd exposure channel without NTA after 3 hours showed that humic substances represented the major fraction of DOC with 48%. Their average molecular weight was 522 g mol^{-1} and aromaticity 2.9 L mg^{-1} m^{-1}, characteristic for fulvic acids. The other fractions were present in lower concentrations, namely neutrals (14%), building blocks (11%), low molecular weight humics (6%) and biopolymers (5%).

3.4.3. Modelled metal species

Modelling (stacked bars in Figure 3.1 a-c and d-f) was performed with the assumption that Fe was present as dissolved Fe(III), since the DGT measurements showed that most Fe was present in labile form. In both Cd exposures, increasing NTA concentrations decreased percentages of free Cd ions (20 nM Cd: 58, 29, 11%, 40 nM Cd: 49, 27, 11%), inorganic complexes and Cd-fulvic acid complexes (20 nM Cd: 7, 5, 3%, 40 nM Cd: 9, 6, 3%) but increased percentages of Cd-NTA complexes (20 nM Cd: 0, 48, 79%, 40 nM Cd: 0, 45, 79%) (Figure 3.1 a). NTA also decreased percentages of free metal ions, inorganic complexes, metal-fulvic acid complexes and increased percentages of metal-NTA complexes for Zn, Cu, Pb and Fe. The modelled Mn speciation was not affected by NTA.

Zn was in all cases the most abundant of the trace metal NTA-species, followed by Fe, Cu and Cd. Pb-NTA and Mn-NTA complexes were present in low concentrations. Ca formed large amounts of complexes with NTA, namely 1.7 µM at the low and 7.5 µM at the high NTA concentration.

Calculated labile Cd concentrations were 151% (20 nM Cd, no NTA), 126% (20 nM Cd, 2×10^{-6} M NTA), 114% (20 nM Cd, 8×10^{-6} M NTA), 141% (40 nM Cd, no M NTA), 140% (40 nM Cd, 8×10^{-6} M NTA) and 90% of DGT-labile Cd concentrations. The best agreement was found at the highest NTA concentrations.

3.4.4. Periphyton characterisation

Chlorophyll-*a* content of periphyton was 4.1 ± 0.3 mg Chl a g dw^{-1} in the control channels and 5.6 ± 1.6 mg Chl a g dw^{-1} in the Cd exposure channels. Semi-quantitative microscopical analysis showed that periphyton was dominated by diatoms (*Achnanthes, Cocconeis, Cymbella, Gomphonema, Navicula, Nitzschia*). Green algae (*Ulothrix*) and Cyanobacteria (*Chamaesiphon, Lyngbya*) were less abundant.

3.4.5. Metal accumulation and speciation

Cd concentrations in control periphyton were 2.3 ± 0.4 nmol Cd g dw^{-1} for total and 1.2 ± 0.1 nmol Cd g dw^{-1} for intracellular content. Cd accumulation in periphyton was influenced by the addition of NTA at both Cd exposure concentrations (Table 3.1). Total and intracellular Cd content was approaching steady-state after 3 hours of exposure, with exception of the two exposures with the highest NTA concentrations. Between 3 and 6 hours total Cd content increased by 56% in the 20 nM Cd exposure, by 48% in the 40 nM Cd exposure and intracellular Cd content by 68% and 62% respectively. After 6 hours of exposure total and intracellular Cd concentrations in periphyton were higher in both Cd exposures with the highest NTA concentration, when compared to the exposures with lower NTA concentrations or no NTA addition.

Chapter 3

	3 hours		6 hours	
Exposure	Total Cd	Intracellular Cd	Total Cd	Intracellular Cd
20 nM Cd / 0 M NTA	5.82 ± 1.01	2.93 ± 0.30	7.59 ± 1.13	3.48 ± 0.29
20 nM Cd / 2x10^{-6} M NTA	7.54 ± 0.59	3.71 ± 0.37	7.34 ± 0.67	3.64 ± 0.08
20 nM Cd / 8x10^{-6} M NTA	6.97 ± 1.11	2.88 ± 0.20	10.88 ± 1.21	4.85 ± 0.51
40 nM Cd / 0 MNTA	14.64 ± 2.74	6.25 ± 0.47	15.74 ± 3.62	5.55 ± 0.78
40 nM Cd / 2x10^{-6} M NTA	15.50 ± 2.06	4.97 ± 0.33	16.54 ± 5.17	5.59 ± 1.40
40 nM Cd / 8x10^{-6} M NTA	12.38 ± 0.78	5.68 ± 1.76	18.35 ± 2.47	9.23 ± 1.81

Table 3.1: Total and intracellular Cd concentrations in periphyton [nmol Cd/ g dw] after 3 and 6 hours of exposure.

Total and intracellular Cd accumulation in periphyton was related to DGT-labile and free Cd concentrations in water (Figure 3.2). A linear relationship between Cd content and DGT-labile Cd concentrations was found after 6 hours of exposure (Figures 3.2 a and a'), whereas no correlation was found in the case of free Cd concentrations (Figures 3.2 b and b'). This indicates that Cd accumulation is controlled by the DGT-labile Cd species. The relationship was less evident after three hours of exposure (data not shown).

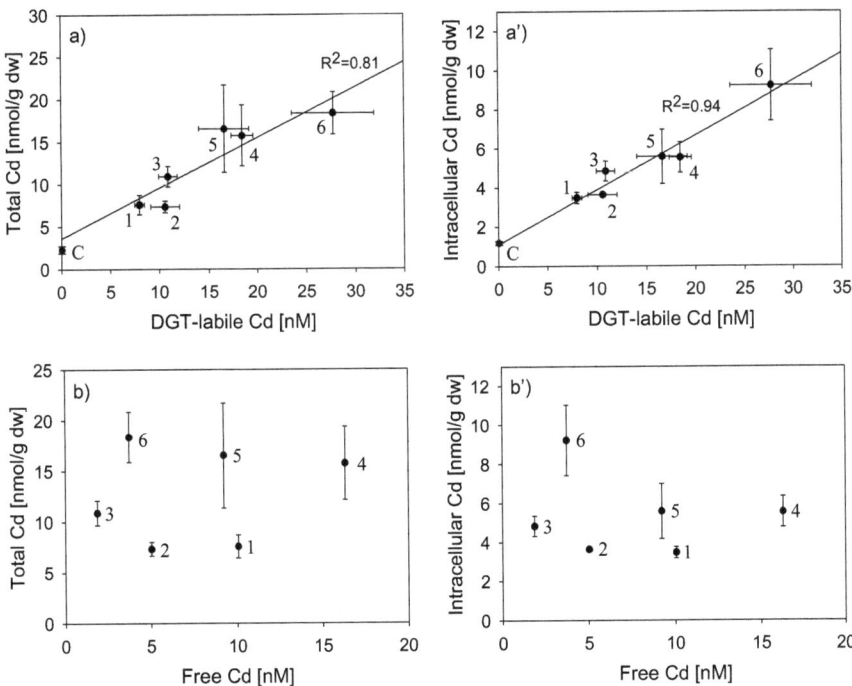

Figure 3.2: Total and intracellular Cd concentrations in periphyton as a function of DGT-labile (a, a') and free Cd (b, b') concentrations in water after six hours of exposure. The numbers represent the different exposures as shown in Figure 3.1 a-c, d-f and C the control. Linear regression (solid line) for Cd content in periphyton and DGT-labile Cd concentrations was performed. Values represent means ± standard deviation. Data points for metal content in periphyton represent three replicates and DGT-labile concentrations four replicates.

3.5. Discussion

3.5.1. Metal speciation in channels

Under constant dissolved Cd concentrations, the addition of NTA changed the Cd speciation at both exposure concentrations by increasing DGT-labile and simultaneously decreasing free Cd concentrations.

DGTs with non-restricted gels measure labile metal species of the dissolved fraction, including free metal ions, inorganic metal complexes, NTA-metal complexes and some fraction of the organic metal complexes [40, 111]. Metals bound to high molecular weight organic ligands, to colloidal metal oxides (Fe, Al, Mn) or to strong binding sites of organic ligands are not measured (non-labile or inert species).

NTA is a strong organic ligand ($\log(K_{CdNTA^-}) = 11.04$) and calculated complexing coefficients for Cd ($\alpha_{CdNTA}=Kx[NTA^{3-}]$) were 4.4 for 2×10^{-6} M NTA and 19 for 8×10^{-6} M NTA. NTA competed for Cd bound to DGT-non-labile ligands (< 0.45µm) and thus increased DGT-labile Cd concentrations at both exposure concentrations. DGT-labile Cd concentrations at the 20 nM Cd exposure did not increase when NTA concentrations were increased from 2×10^{-6} M to 8×10^{-6} M. This indicates that the remaining non-labile Cd was bound to ligands with complexing coefficients higher than 19 or to slowly exchanging colloidal species. The dissociation of metals from particulate complexes (>0.45µm, non-labile) due to NTA addition was observed for Fe, where dissolved concentrations increased with increasing NTA concentrations.

NTA did not increase DGT-labile concentrations of Zn and Mn since both metals were mostly present in labile form at both Cd concentrations without NTA. A high proportion of Cu (~49%, ($\log(K_{CuNTA^-}) = 14.28$)) was bound to non-labile ligands, which did not exchange with NTA. Complexing coefficients (α_{CuNTA}) were 7.7×10^{3} for 2×10^{-6} M NTA and 3.4×10^{4} for 8×10^{-6} M NTA. Cu was thus bound to natural ligands with complexing coefficients higher than 3.4×10^{4} or to slowly exchanging colloidal species.

The addition of NTA decreased modelled free Cd concentrations due to complexation with NTA. Free Cd concentrations might be overestimated at each exposure, since other organic ligands were identified with LC-OCD but were not included in the model. Also Cd fulvic acid complexes decreased upon NTA addition, because they have lower stability constants than Cd-NTA complexes [28].

With increasing NTA concentrations, calculated labile Cd concentrations agreed better with DGT-labile concentrations at both Cd exposures. Since non-labile ligands are not included in the model, but decrease the measured DGT-labile species, calculated labile Cd concentrations are higher than DGT-labile concentrations without the addition of NTA. The increase of labile species with NTA addition leads to a better agreement between calculated labile Cd and measured DGT-labile Cd.

3.5.2. Accumulation and speciation

Total and intracellular Cd content of periphyton increased in the presence of NTA and could be related to Cd speciation in water (Figure 3.2). The relationship was more evident after six hours of exposure, which may be due to the longer exposure times of DGT devices and periphyton and therefore more precise data. NTA had an effect on the kinetics of Cd uptake. Cd content in periphyton increased at the highest NTA concentration in both Cd exposures between three and six hours (20 nM: total 56%, intracellular 68%; 40 nM: total 48%, intracellular 62%), whereas it remained fairly constant at the other exposures (Table 3.1). This might be due to a slower diffusion of Cd-NTA complexes through the matrix of periphyton than diffusion of free Cd ions or inorganic complexes. It has been shown for diffusion layers of DGTs, which might be comparable to the matrix, that the diffusion coefficient of Cd-NTA complexes is 70% of that of the free metal ion [111]. A diffusion limitation of metals by the matrix of periphyton was also suggested when Cd uptake kinetics of algae and periphyton were compared [13].

The increase of total and intracellular Cd content in periphyton could be related to the increase in DGT-labile Cd concentrations in water (Figure 3.2). The modelling results show that DGT-labile concentrations should be dominated by Cd-NTA complexes. Since modelled free Cd concentrations decreased with NTA addition, the increase in Cd content appears to be related to the diffusion of Cd-NTA complexes.

Our results do not agree with equilibrium based models, i.e. the FIAM and BLM. They assume equilibrium between metal species in the bulk solution and metals bound to transport sites of the plasma membrane. This implies that the internalization flux of metals across the cell membrane (J_{int}) is rate-limiting, meaning that the diffusion flux of metal species from the bulk solution to the plasma membrane (J_{diff}) is faster than J_{int}. Non-equilibrium based models consider diffusion fluxes of metal species and internalization fluxes. Labile metal species can control bioaccumulation in the case of diffusion limitation of free metal ions, which has been shown theoretically [23, 82] and in experiments with algae [85] and periphyton [86]. The present results, as well as a previous study [13] have shown that the matrix of periphyton influences the diffusion of metals to algae.

Our results confirm that dynamic metal species can control bioaccumulation and that they need to be included in models which predict bioavailability, especially in natural aquatic systems, where chemical equilibrium is almost never achieved. Various dynamic speciation sensors are available nowadays, which could help to interpret the results obtained in natural waters [23].

3.5.3. Environmental relevance

Our study shows that an anthropogenic ligand can increase the bioavailability of a non-essential metal by dissociation from non-labile complexes, which would be normally not bioavailable to aquatic organisms. Concentrations of NTA in surface water range from 2 to 63 nM [114, 115] and have been found to be as high as 157 nM in drinking-water and raw water samples [116]. NTA may not play an important role in changing Cd speciation in natural waters, since environmental concentrations are much lower than in our study and NTA is rapidly degraded. The question remains if other, more stable organic ligands introduced into the aquatic environment by human activities could increase concentrations of labile complexes of non-essential and essential metals thus increasing their bioavailability and toxicity. Natural small organic ligands have been identified in this study by LC-OCD. Our results indicate that labile metal complexes can contribute to the metal flux towards algae of periphyton.

Chapter 4

Cadmium speciation and accumulation in periphyton in a small stream during rain events

This chapter is published in *Environmental pollution, 2010, 158: 641–648*

4.1. Abstract

The bioavailability of metals to aquatic organisms has been extensively studied with algae in defined media, but only a few studies are available for periphyton in natural freshwaters. Field investigations are needed to validate data from laboratory experiments, since natural aquatic systems show dynamic variations in chemical composition, and periphyton is a community of various algal species. The objective of the present study was to investigate the accumulation of Cd in periphyton under field conditions as a function of dynamic variations of Cd speciation in water during rain events. Speciation in water was determined *in situ* with diffusion gradient in thin-films (DGT). Concentrations of major Cd species were estimated by modelling and compared with DGT measurements, by including diffusion coefficients of humic substances. During the rain events dissolved Cd concentrations increased from 0.17 nM to concentrations between 0.27 and 0.36 nM and free Cd concentrations from 0.10 nM to concentrations between 0.12 and 0.14 nM. DGT-labile Cd concentrations did not change during the rain events and were between 70 and 97% of total dissolved Cd. They were in good agreement with average concentrations of calculated labile Cd, whose concentrations varied during the rain events. Periphyton responded sensitively to these changes despite higher concentrations of Zn and Mn. Total Cd content varied from 1.70 ± 0.2 nmol Cd/g dw to 2.8 ± 0.7 nmol Cd/g dw and intracellular content from 1.30 ± 0.2 nmol Cd/g dw to 2.0 ± 0.1 nmol Cd/g dw. Cd accumulation in periphyton might be controlled by either free or DGT-labile Cd concentrations. After the rain events, when dissolved Cd concentrations were decreasing Cd content in periphyton showed different courses, which might be explained by intracellular metal regulation mechanisms. As for Cd, the concentrations of other metals (Zn, Mn, Cu, Pb, Fe) in periphyton were also following the dynamic variations of metal concentrations in water.

4.2. Introduction

Essential (e.g. Cu, Zn) as well as non-essential metals (e.g. Cd, Pb, Hg) are elevated in aquatic systems due to human activities [3-6]. Their bioavailability and hence uptake and effects on aquatic organisms depend on metal speciation, which is determined by the composition of the natural water (inorganic and organic ligands, pH) [16]. Metal uptake by algae in defined media has been successfully predicted by the free ion activity model (FIAM) and the derived biotic ligand model (BLM), although some exceptions have been found [64, 77, 78]. Both are equilibrium-based models, not considering the dynamic features (mobility, lability) of metal complexes. Non-equilibrium based models consider diffusion and internalization fluxes [23, 82, 83]. Diffusion of free metal ions from the bulk solution to the plasma membrane can be limiting, in which case metal uptake will be controlled by the concentrations of chemically labile species [82, 84].

Since natural freshwaters are temporally and spatially heterogeneous environments in terms of physical properties and chemical as well as biological composition, aquatic organisms will be exposed to different metal species in space and time. Periphyton, a natural assemblage of heterotrophic and autotrophic organisms, is the most important primary producer in running waters [55]. Elevated Cd concentrations in water were shown to impact various endpoints in periphyton, including biomass related parameters (chlorophyll a, dry weight) [60, 61], settlement, development and species composition [62]. Cd concentrations in periphyton were found to be higher in streams impacted by metal pollution [3-5], when compared to relatively unpolluted sites [5, 59]. Although several laboratory [13, 99-101] and field studies [59, 92] investigated the accumulation of Cd in periphyton, only one study related metal accumulation (Cu, Zn) in periphyton to metal speciation in natural freshwater [86].

Various techniques for speciation analysis in natural waters are available [12, 26]. Metal species, namely free ions or dynamic species, are measured either directly in the field (*in situ*) or on discrete water samples in the laboratory (*ex situ*). Diffusion gradient in thin-films (DGT) is an *in situ* technique [40, 102], giving mean labile metal concentrations during the time of deployment to which also aquatic organism are exposed. In addition, chemical speciation programs (WHAM 6 [27], visual

MINTEQ [29] using different models [27, 30, 31] for the metal complexation by humic and fulvic acids are used to estimate the concentrations of the different metal species. But in natural waters, where many components are unknown and where other unidentified organic ligands are present, such calculations have to be compared with results obtained by speciation techniques.

The aim of the present study was to investigate the accumulation of Cd in periphyton under field conditions in dependence of dynamic variations of Cd speciation in water during rain events. Speciation was measured *in situ* with diffusion gradient in thin-films (DGT) and major Cd species were estimated by modelling the complexation with inorganic ligands and fulvic acids. Concentrations and characteristics of humic substances for the model were determined. Accumulation of other metals (Zn, Mn, Cu, Pb, Fe) in periphyton and their speciation in water were also determined in order to explain Cd accumulation and speciation data.

4.3. Materials and methods

4.3.1. Site description

The small stream Altbach (canton Zurich, Switzerland) flows through several towns and agricultural areas and receives effluents from a sewage treatment plant. Stream Altbach at the experimental site is around 2.4 m wide, 25-30 cm deep with the streambed consisting of stones and sediment. At a short distance downstream, Altbach flows together with another small stream to form the stream Chriesbach. Dissolved Cd concentrations in stream Altbach during a preliminary field campaign were 0.09 nM and increased to 0.35 nM during a rain event.

4.3.2. Periphyton colonization and translocation

Periphyton was precolonised on glass slides in two artificial flow-through channels [103]. Details about the channel setup can be found in a previous publication [13]. The channels were supplied with natural freshwater from the nearby Chriesbach stream containing low dissolved Cd background concentrations and the organisms for the colonization of periphyton. Water flow in the channels was 10 L min^{-1}.

Illumination was provided by two PAR-lamps (Osram HQL (MBF-U) Deluxe 400 W) overhanging the channels. The photoperiod was 15 h darkness, 9 h light and the average light intensity 298 ± 51 μE m^{-2} sec^{-1}.

After three weeks of colonization in the channels sufficient biomass and a homogeneous coverage of slides with periphyton were available. For the transport to the field site, five glass slides were fixed on both sides of Teflon racks (12 cm x 20 cm) and put in a plastic box containing Chriesbach water. Racks were fixed in four rows, vertically and parallel to the water flow in stream Altbach about 6 cm below the water surface on iron bars.

The time point of translocation was chosen, in order to allow for acclimatization of periphyton to the Altbach water before the start of the rain events (34 hours).

4.3.3. Field study

Periphyton slides for total and intracellular metal concentrations were sampled once during colonization, 28.5 hours after translocation (prior to the first rain event), twice a day during three days of rain and three times after the rain events. Water samples for major cation and anion concentrations and for characterization of natural organic matter were collected at the same times. Water samples for dissolved metal concentrations were collected manually three times during colonization, twice prior to the first rain event and afterwards every hour by an autosampling device (ISCO 6700, Isco, Inc.). Two DGT devices were fixed on both sides of U-shaped polymethyl methacrylate holders and screwed in tight to the Teflon racks. Thus DGT devices were placed close to colonized periphyton slides and parallel to the water flow. They were deployed for around 24 hours in a sequence, so that the deployment time was overlapped by two others by 7 and 17 hours. Temperature and pH were measured automatically every 30 min during colonization and the field study by a universal pocket meter (Multi 340i, Metrohm). Rainfall [mm], with a resolution of 10 min, was obtained from a meteorological station located 1.4 km away from the field site.

4.3.4. Clean trace metal handling

To avoid contamination of samples and equipment, plastic gloves (Semadeni) were used for all procedures and the material for the field was sealed in plastic bags. Components of the automatic sampler, boxes for periphyton transport, Teflon racks, polypropylene vials, bottles and beakers, DGT devices, holders and filters, syringes and filtration units were placed in 0.1 M HNO_3 for at least 24 h and then properly rinsed with nanopure water. Cellulose nitrate filters for metal content in periphyton were boiled twice in 0.1 M HNO_3, rinsed with nanopure water and dried twice at 50°C for 15 hours in an oven. Except for the sampling at the field site all handling was performed in a clean bench.

4.3.5. Water sampling for dissolved metal concentrations

Manual water samples (14 ml) for dissolved metal concentrations were filtered on site (0.45 µm filters, Milipore) into polypropylene tubes using a plastic syringe (BD Plastipak, 50 ml). Filters and syringes were previously thoroughly rinsed with water from the channel or stream. Water samples from the autosampling device (1 L) were transported to the laboratory, thoroughly shaken to resuspend sediments and filtered in the same way. All samples were acidified to 0.24 M with HNO_3 (65% suprapure. Merck) and kept at 4°C in the dark till analysis. Dissolved metal concentrations for the time points of periphyton sampling were extrapolated from the concentrations of the autosampling device.

4.3.6. NOM analysis and modelling of metal speciation

Natural organic matter was analyzed using size-exclusion chromatography in connection to on-line, high sensitivity organic carbon detection (LC-OCD) [32-34]. Fractions of hydrophilic dissolved organic carbon determined by this method are biopolymers (polysaccharides, proteins), humic substances (humic and fulvic acids), degradation products of humic substances (building blocks), low molecular weight humics (degradation products of humic substances), low molecular weight acids (mono- and dicarboxylic acids) and neutrals/amphiphilics (amino acids, alcohols,

Chapter 4

aldehydes, ketones). Characterisation of molecular weight and aromaticity of humic substances allows further differentiation between fulvic and humic acids. Concentrations of characterised humic substances were used together with dissolved concentrations of metals (Cd, Zn, Mn, Cu, Pb, Fe), alkalinity, cations (Na^+, K^+, Ca^{2+}, Mg^{2+}) and anions (Cl^-, SO_4^{2-}, NO_3^-, o-PO_4^{3-}), as well as pH and temperature to model concentrations of free metal ions, inorganic metal complexes and metal-fulvic acid complexes using the program vMINTEQ [29] with the Stockholm humic acid model [31]. Calculations were performed with the assumptions of Fe being present as dissolved Fe(III) or precipitated as Fe(III) oxides. DGT-labile metal concentrations were compared with calculated labile metal concentrations (Me-labile$_{calc}$) by using:

Me-labile$_{calc}$ = Σ free metal ions + inorganic complexes + 0.2 x (metal-FA complexes) (1)

considering that fulvic acids have diffusion coefficients that are typically 20% of that of the free metal ion [112].

4.3.7. Preparation, sampling and processing of DGT devices

Labile metal concentrations in water were measured with DGT (Diffusion gradients in thin-films). The procedure for making DGT devices followed the recommendation by Zhang and Davison [102] and is described in detail in Bradac et al. [13]. DGT holders were screwed in tight to the Teflon racks before the start of the rain events. They were replaced after a deployment time of about 24 hours. For measurement of labile metal concentrations, the resin gel layer was removed and placed for 24 h in a 14 mL polypropylene vial containing 2 mL of 1.66 M HNO_3 (65% suprapure, Merck) and then diluted sevenfold. To calculate DGT-labile metal concentrations in water, average water temperatures during the time of deployment were used to obtain diffusion coefficients for free metal ions. The calculations for labile metal concentrations are described by Zhang and Davison [39] and the diffusion coefficients for free metal ions were measured by Hao Zhang, DGT Research Ltd.,

Lancaster (personal communication). Percentages of DGT-labile metal concentrations were related to averaged total dissolved metal concentrations over the time of DGT deployment. Values of DGT replicates were compared to mean dissolved concentrations during the time of deployment and were excluded from the results if they were larger than means plus one standard deviation of dissolved metal concentrations. Measurements with only one replicate were also excluded.

4.3.8. Periphyton sampling and processing

Six slides from the colonisation channel and one rack (10 slides) at each time point during the field study were sampled for metal analysis of periphyton. To avoid dehydration and contamination of periphyton, racks (or slides) were stored in a plastic box containing a small amount of stream water till further processing. Periphyton was analysed for total and intracellular metal content using EDTA for the removal of surface-bound metals. The detailed processing of periphyton samples is described in Bradac et al. [13]. The measured metal content in periphyton was related to the measured dry weight.

Chlorophyll *a* was extracted from 5 mL of the suspension with ethanol [113] and concentrations were measured using the HPLC method described by Murray et al. [104]. For determination of species composition 5 mL from periphyton suspensions were fixed with 4% formaldehyde. It was determined semi-quantitatively, using an inverted phase contrast microscope with a magnification of 640.

4.3.9. Metal analysis

Metal concentrations were measured with high resolution inductively coupled plasma mass spectrometry (HR-ICP-MS) (Element 2, Thermo Finnigan). The accuracy of the ICP-MS measurements was checked using the reference waters SLRS-4 (National Research Council Canada, errors: Cd < 11%, Zn < 8 %, Mn < 9%, Cu < 9 %, Pb < 12 %, Fe < 13 %) and TM-28.2 (National Research Council Canada, errors: Cd < 8 %, Zn < 10 %, Mn < 8 %, Cu < 7 %, Pb < 10 %, Fe < 11 %) as well as plankton reference material (CRM 414, Institute for reference materials and measurements,

European Commission, Belgium, errors: Cd < 9 %, Zn < 8 %, Mn < 11 %, Cu < 8 %, Pb < 11 %, Fe < 12 %).

4.3.10. Water chemistry

Water samples for DOC, alkalinity, major cation and anion analysis were taken at the same time as periphyton. Concentrations of major cations were determined by ICP-OES and anion concentrations were measured by ion chromatography (Metrohm). Alkalinity measurements were performed by titration (with HCl 0.1M until pH 4.5).

4.4. Results

4.4.1. Dissolved metal concentrations

The first rain event started 1.4 days after translocation of periphyton, with a rain amount of 2.7 mm within 15.2 hours. The second and third rain event lasted for 17.3 and 17.2 hours with a rain amount of 4.0 and 12.2 mm respectively (Figures 4.1 a-f). The three rain events increased dissolved concentrations of all metals, but the dynamics differed from metal to metal. Dissolved metal concentrations during the different phases of the field study are summarized in Table 4.1. With the exception of zinc, concentrations of all metals in stream Altbach prior to the first rain event were higher than in the colonisation channels. Dissolved Cd concentrations were 0.11 ± 0.01 nM in the colonisation channels, 0.17 ± 0.02 nM prior to the first rain event, increased to 0.27 to 0.36 nM during the rain events and decreased to 0.055 ± 0.016 nM after the last rain event. Dissolved concentrations of Zn and Mn in water of stream Altbach were much higher when compared to Cd. Ratios of Zn : Cd were $2.3 \times 10^3 \pm 1.1 \times 10^3$ during the field study, with some higher ratios between 4.0×10^3 and 8.5×10^3 prior to the maximum of the second rain event. Ratios of Mn : Cd were $1.0 \times 10^3 \pm 0.3 \times 10^3$ till the end of the last rain event and increased to $3.1 \times 10^3 \pm 0.9 \times 10^3$ for the rest of the field study.

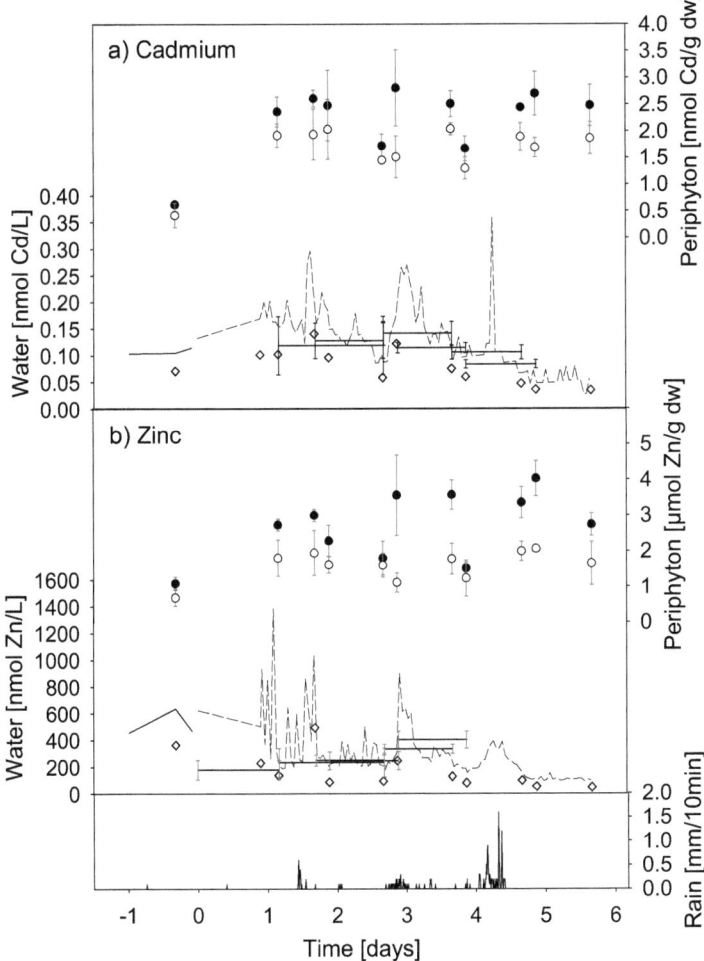

Figure 4.1 a, b: Metal speciation in water and metal content in periphyton for cadmium and zinc. Total (black points) and intracellular (white points) metal concentrations in periphyton as well as dissolved DGT labile and free metal concentrations in water. Solid curves represent dissolved metal concentrations during colonisation, dashed curves dissolved metal concentrations in the stream Altbach, horizontal lines DGT labile metal concentrations and diamonds free metal concentrations modelled without Fe(III). The three rain events are shown as rain amounts [mm/10min] at the bottom of both graphs. Time 0 is the point of translocation. Values represent means ± standard deviation. Data points for metal content in periphyton represent three replicates and DGT labile concentrations two to four replicates.

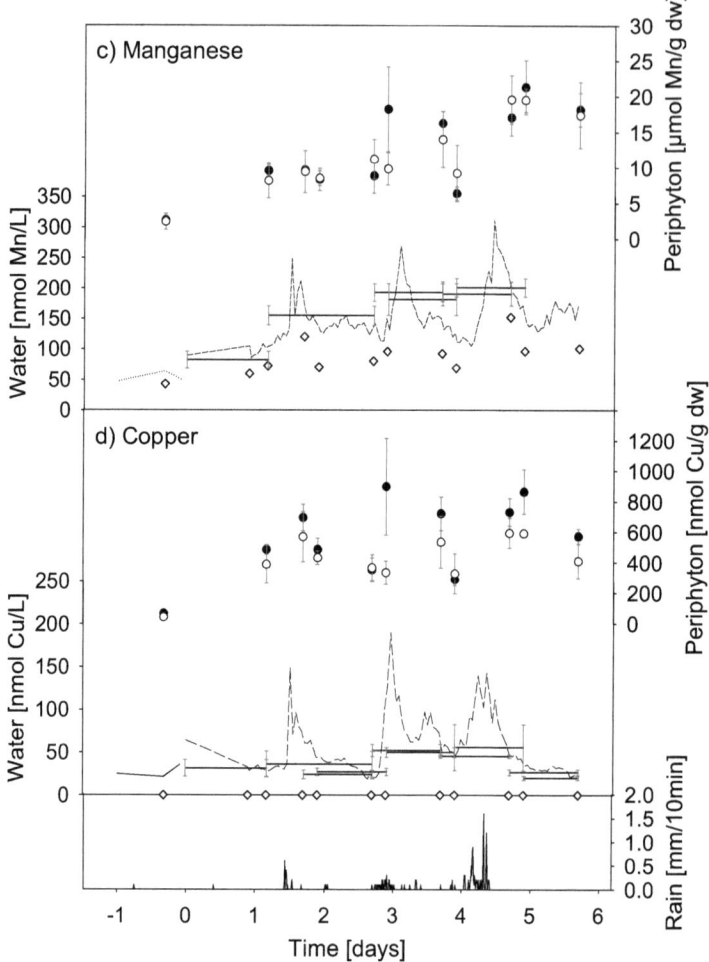

Figure 4.1 c, d: Metal speciation in water and metal content in periphyton for manganese and copper. The different symbols and lines are described in Figure 4.1 a, b.

Chapter 4

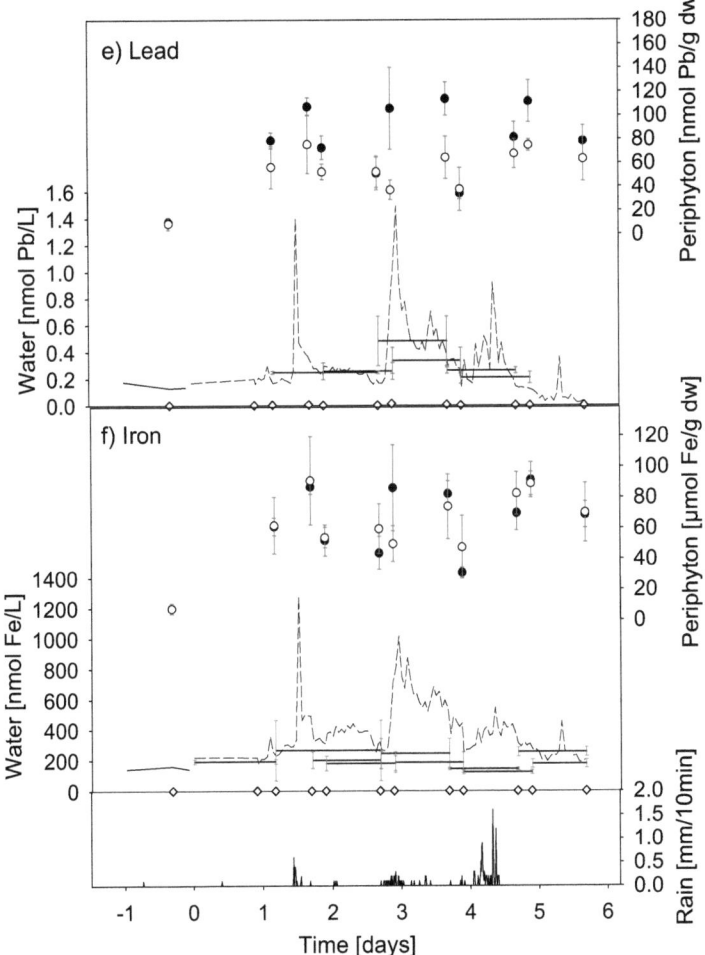

Figure 4.1 e, f: Metal speciation in water and metal content in periphyton for lead and iron. The different symbols and lines are described in Figure 4.1 a, b.

During the field study alkalinity was $5.59 \times 10^{-3} \pm 0.54 \times 10^{-3}$ M, chloride $1.48 \times 10^{-3} \pm 0.47 \times 10^{-3}$ M, calcium $2.46 \times 10^{-3} \pm 0.23 \times 10^{-3}$ M, pH 7.9 ± 0.13 and temperature 11 ± 0.7 °C.

Phase of the field study	Cd [nM]	Zn [nM]	Mn [nM]	Cu [nM]	Pb [nM]	Fe [nM]
Colonisation	0.11 ± 0.01	520 ± 101	54 ± 9	27 ± 8	0.15 ± 0.02	148 ± 11
Prior to 1st rain event	0.17 ± 0.02	486 ± 339	107 ± 14	34 ± 9	0.20 ± 0.04	258 ± 51
1st rain event (I)	0.30	242	249	149	1.4	1281
2nd rain event (II)	0.27	901	269	191	1.5	1015
3rd rain event (III)	0.36	392	311	143	0.92	543
After last rain event	0.055 ± 0.016	105 ± 7	162 ± 13	29 ± 4	0.073 ± 0.089	248 ± 67

Table 4.1: Dissolved metal concentrations during colonisation, prior to the first rain event, highest concentrations during rain events (I - III) and after the last rain event (0.5 days before the end of the field study). Values are means ± standard deviations from manual samples and the automatic sampler. Means represent 3 samples for the colonisation, 16 for the time prior to the first rain event and 13 for the time after the last rain event.

4.4.2. DGT-labile metal concentrations

Concentrations of DGT-labile metals are shown in Figure 4.1 a-f. The increases in dissolved metal concentrations during the rain events did not consequently increase DGT-labile metal concentrations. Correlations of DGT-labile metal concentrations and average dissolved metal concentrations during the time of DGT deployment are represented as R^2 in the text and DGT-labile metals as % of total dissolved concentrations. Despite the low dissolved Cd concentrations in stream Altbach, DGT-labile Cd concentrations could be measured and were between 0.08 and 0.14 nM (70 to 97%). A slight increase in DGT-labile Cd concentrations from 0.13 nM to 0.14 nM was measured only during the first rain event (R^2=0.66). DGT-labile Zn concentrations were 179 nM prior to the first rain event, increased to 243 ± 9 nM during the first till the beginning of the second rain event and were 334 nM during the second rain event (R^2=0.15, 27 and 93%). Concentrations of DGT-labile Mn prior to the first rain event were 82 nM and increased continuously from 155 nM to 201 nM with the rain events (R^2=0.98, 86 to 117%). DGT-labile Cu concentrations prior to the first, between the first and second and after the third rain event were between 20 and 32 nM and increased during the rain events to concentrations between 36 and 56 nM (R^2=0.90, 57 to 88%). Increases of DGT-labile Pb concentrations were measured during the second rain event (0.41 ± 0.10 nM), whereas other concentrations were 0.24 ± 0.20 nM (R^2=0.74, 27 to 93%). Increases of DGT-labile Fe concentrations were measured during the first and second rain event, as well as at the end of the study (256 ± 11 nM), whereas other concentrations were 173 ± 28 nM (R^2=0.01, 30-101%).

4.4.3. NOM characterisation

Chromatographic fractions of hydrophilic dissolved organic carbon (DOC) measured by LC-OCD are shown in Table 4.2. Humic substances represented the major fraction with 44% of DOC in the colonisation channels and 30 to 46% in the stream Altbach. Average molecular weights were 542 to 1002 g mol^{-1} and aromaticity 2.7 to 3.5 L mg^{-1} m^{-1}, characteristic for fulvic acids. Molecular weights of fulvic acids from samples taken during the rain events were 1.4 to 1.8 larger than from samples taken prior to the rain events. Fulvic acids with similar aromaticity and molecular weights were determined in effluents from sewage treatment plants (Isabelle Worms, EPFL, Switzerland, personal communication). The other fractions were present in lower concentrations, namely building blocks (12 - 19%), neutrals (12 - 16%), low molecular weight (LMW) humics (5 - 7%), biopolymers (4 - 6%) and low molecular weight (LMW) acids (0.1 - 1.9%). The three rain events caused very dynamic concentration changes of DOC and humic substances with highest concentrations during the first and second and after the third rain event.

Time [days]	DOC	Humic substances	Building blocks	Neutrals	LMW humics	Biopolymers	LMW acids
-0.3	1710	759 (44%)	266	223	94	65	5
0.9	2146	742 (35%)	401	330	124	91	40
1.2	2254	919 (41%)	279	337	140	145	0
1.7 (I)	3230	1022 (32%)	575	493	168	182	42
1.9	2413	806 (33%)	411	328	141	118	27
2.7	1800	694 (39%)	273	259	118	86	1
2.9 (II)	3282	982 (30%)	585	524	185	166	31
3.7	2299	804 (35%)	385	336	143	111	22
3.9	2265	785 (35%)	373	345	142	104	19
4.7 (III)	2374	995 (42%)	289	313	133	130	16
4.9	1739	793 (46%)	254	216	94	73	11
5.7	1359	627 (46%)	219	183	78	52	6

Table 4.2: Concentrations [μg C L^{-1}] of dissolved organic carbon (DOC) and of the identified hydrophilic chromatographic fractions obtained by LC-OCD. Values in brackets are percentages of dissolved organic carbon and Roman numbers (I-III) indicate maximum concentrations during rain events. Time -0.3 days is the sampling time during colonisation, the other samples were taken during the field study.

4.4.4. Modelled metal species

Results of major metal species without including Fe(III) in the model are shown in Figure 4.2 a-e. Percentages of free metal ion concentrations in stream Altbach were high for Cd (56-70%), Zn (40-63%) and Mn (52-76%) but low for Cu (0.03-0.09%) and Pb (0.9-1.8%). Percentages of metal-fulvic acid complexes were high for Cu (95-98%) and Pb (39-63%) but low for Cd (6-11%), Zn (7-11%) and Mn (0.05-0.13%). Of the total metal-fulvic acid complexes, the dominant ones were copper and zinc. The inclusion of dissolved Fe in the model, assumed to be present as Fe(III), increased concentrations of calculated free metal ions more in the case of Cu (0.13-0.41%) and Pb (1.2-3%), and less in the case of Cd (60-76%) and Zn (42-67%) and did not affect Mn. Percentages of metal-fulvic acid complexes were strongly decreased for Cd (0.001-0.004%), Zn (3.1-4.4%) and Pb (21-38%), moderately for Cu (77-89%) and did not change for Mn. Dominant metals bound to fulvic acids were iron, copper and zinc. In this case, Fe(III) was present as inorganic complexes by 14-33% and as metal-fulvic acid complexes between 67 and 86%. All modelling results are available in the supporting information (S1).

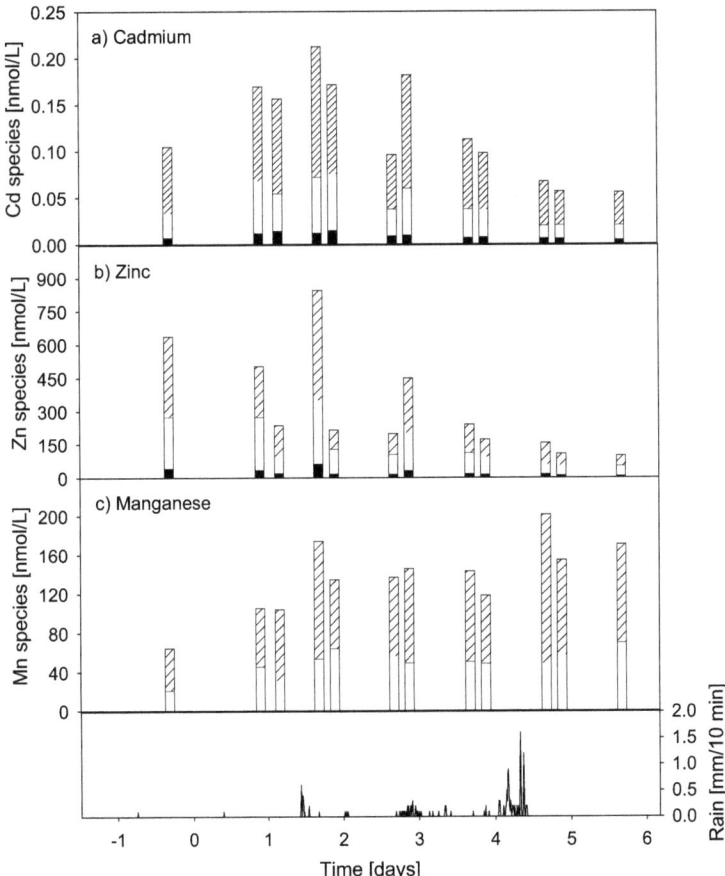

Figure 4.2 a-c: Modelled metal species for cadmium, zinc and manganese. Concentrations of free metal ions (diagonal texture), inorganic (white texture) and fulvic acid metal complexes (black texture) modelled with vMINTEQ. Modelling was performed without Fe(III) and with fulvic acid concentrations determined with LC-OCD. Time point 0 represents the time of translocation.

Chapter 4

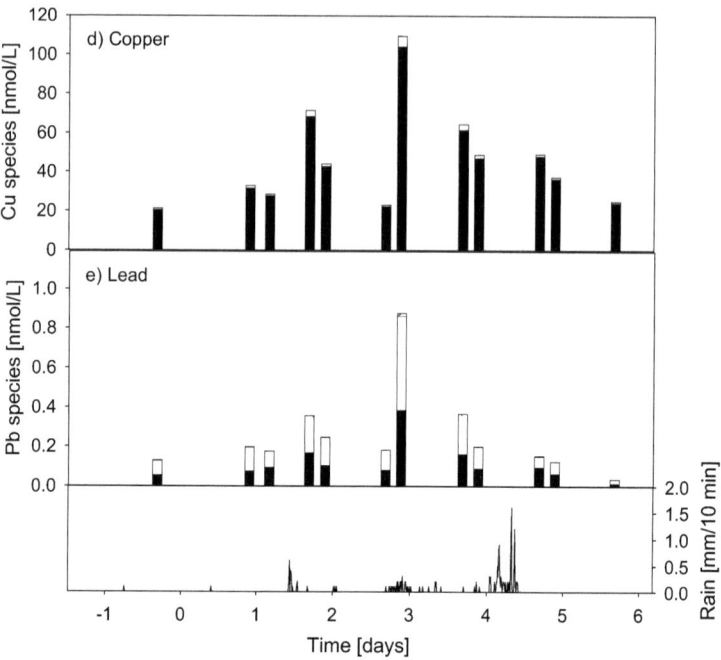

Figure 4.2 d-e: Modelled metal species for copper and lead. Textures of metal species are described in Figure 4.2 a-c.

The free Cd concentration (calculated without Fe(III)) in the colonisation channels was 0.07 nM and increased after translocation to 0.10 nM prior to the first rain event. During the first and second rain event concentrations increased to 0.14 and 0.12 nM respectively and decreased then continuously from 0.07 to 0.03 nM.

Calculated labile metal concentrations were close to measured DGT-labile concentrations for Cd and Mn, with calculated concentrations being 82 to 125% and 73 to 127% of DGT-labile concentrations respectively. The variability of differences was higher for Zn (67 to 193%) and Pb (44 to 110%). Calculated labile Cu concentrations were all lower than DGT-labile concentrations (18-50%).

4.4.5. Periphyton characterisation

Chlorophyll-*a* content of periphyton in the colonisation channels was 4.8 ± 0.5 mg Chl a g dw^{-1} and average concentration during the field study was 6.7 ± 1 mg Chl a g dw^{-1}, indicating a constant biomass. Semi-quantitative microscopical analysis showed that periphyton was dominated by diatoms (*Achnanthes, Cocconeis, Cymbella, Fragilaria, Gomphonema, Navicula, Nitzschia, Synedra, Tabellaria*). Green algae (*Ulothrix, Gongrosira*) and Cyanobacteria (*Pseudoanabena, Phormidium, Planktothrix, Chroococcus*) were less abundant. Composition and abundance of algae in periphyton was not affected by translocation and exposure in the field.

4.4.6. Metal accumulation in periphyton

Total Cd content in colonised periphyton before translocation was 0.62 ± 0.02 nmol Cd g dw^{-1} and intracellular 0.42 ± 0.22 nmol Cd g dw^{-1}. Despite the only slightly higher dissolved Cd concentrations in stream Altbach prior to the first rain event (factor 1.6) the total content increased to 2.4 ± 0.3 nmol Cd g dw^{-1} (factor 3.8) and the intracellular content to 1.9 ± 0.2 nmol Cd g dw^{-1} (Factor 4.6) after translocation. Total and intracellular Cd concentrations in periphyton were very dynamic during the three rain events, following the concentrations of dissolved Cd in water (Figure 4.1 a). Total and intracellular concentrations varied by a factor of 1.7 and 1.6 respectively, in contrast to dissolved Cd concentrations, which varied by a factor of 13. Changes of Cd concentrations in periphyton differed during each rain event, with no changes during the first rain event (2.5 ± 0.1 nmol Cd g dw^{-1} for total and 2.0 ± 0.1 nmol Cd g dw^{-1} for intracellular content). In contrast, during the second rain event, intracellular Cd concentrations in periphyton increased over 24 h from 1.5 nmol Cd g dw^{-1} to 2.0 ± 0.1 nmol Cd g dw^{-1}, whereas total concentrations increased first from 1.7 ± 0.2 nmol Cd g dw^{-1} to 2.8 ± 0.7 nmol Cd g dw^{-1} and then decreased slightly to 2.5 ± 0.2 nmol Cd g dw^{-1}. The increase during the third rain event was from 1.7 ± 0.2 nmol Cd g dw^{-1} to 2.4 nmol Cd g dw^{-1} for total and from 1.3 ± 0.2 nmol Cd g dw^{-1} to 1.9 ± 0.3 nmol Cd g dw^{-1} for intracellular content. Uptake rate constants obtained in previous microcosm experiments with periphyton [13] predicted fairly well the increases in intracellular Cd concentrations during the second and third rain event with predicted concentrations being 69 to 97% of measured values.

Different courses of Cd concentrations in periphyton were also observed after the end of the rain events, which were accompanied by decreases of dissolved Cd concentrations in water. Total and intracellular Cd concentrations in periphyton decreased by a factor of 1.5 after the first and second rain event, the decrease being faster after the second rain event. Despite the continuous decrease in dissolved Cd concentrations in water after the third rain event, total and intracellular Cd concentrations in periphyton remained constant and were 2.5 ± 0.1 nmol Cd g dw^{-1} for total and 1.8 ± 0.1 nmol Cd g dw^{-1} for intracellular content, similar to the concentrations prior to the first rain event.

As for Cd, the concentrations of the other metals in periphyton were also following the dissolved metal concentrations in water (Figures 4.1 b-f). Translocation increased total and intracellular contents of all metals in periphyton. Total and intracellular concentrations of Zn and Mn did not change during the first rain event, whereas content of Cu, Pb and Fe increased. Zn, Mn, Cu, Pb and Fe content of periphyton increased also during the second and third rain event, showing a similar fast increase as Cd in total concentrations during the second rain event. After the first rain event metal content of periphyton was nearly constant for Zn, Pb, Fe, increased for Mn and decreased for Cu. Zn, Mn, Cu, Pb and Fe content of periphyton decreased after the second rain event as fast as for Cd. After the third rain event metal concentrations in periphyton were nearly constant for Zn, Mn, Pb and Fe, but decreased for Cu. As dissolved Mn concentrations in water increased steadily during the rain events, so did the Mn content in periphyton and was higher than at the beginning of the field study. Values of total and intracellular metal contents are available in the supporting information (S2).

4.5. Discussion

4.5.1. Metal speciation

The variations of rainfall are reflected in the dissolved metal concentrations in stream Altbach. Possible sources of Zn and Cu might be release of metals from resuspended contaminated sediments, runoff, wet deposition and discharge from the sewage treatment plant located 2.9 km upstream. In the case of Mn and Fe the main source might be the resuspension and dissolution from sediments. With respect to the water quality criteria of Switzerland, dissolved concentrations were not exceeded for Cd (0.44 nM) but for Cu (31 nM) and Zn (76 nM) [117].

Using DGT with non-restricted gels allows the determination of labile metal species of the dissolved fraction, which are free metal ions, inorganic metal complexes and some fraction of the organic metal complexes [40]. Metals bound to high molecular weight organic ligands, to colloidal metal oxides (Fe, Al, Mn) or to strong binding sites of organic ligands are not measured (non-labile or inert species).

The high percentages of DGT-labile Cd (70-97%) measured in stream Altbach indicate that only a small proportion of Cd is present as non-labile species. Percentages of DGT-labile Cd were similar in another freshwater stream in the same region, with similar pH, dissolved Cd and DOC concentrations, but lower (36%) in a softwater river in England with a high humic acid content [26]. Due to higher dissolved Cd concentrations and a lower pH, DGT-labile Cd concentrations in two Australian rivers were equal to the dissolved concentrations [118]. Most important factors that influence DGT-labile metal speciation are concentration and nature (complexation properties) of dissolved organic carbon, pH, inorganic colloids and concentration of other metals with different stability constants for natural organic matter.

DGT measurements in our study were in good agreement with concentrations of modelled labile Cd species, showing a high percentage of Cd being present as free Cd ions (56-70%). In contrast, percentages of free Cd concentrations in other streams with similar dissolved Cd concentrations determined with *ex situ* [11] or *in situ* [26] speciation techniques were between 2 and 15% of dissolved Cd. Modelled free Cd concentrations might be overestimated in our study, since other organic ligands identified with LC-OCD were present in similar concentrations in the water as fulvic acids, but were not included in the model.

The inclusion of Fe(III) in the model slightly increased free Cd concentrations, due to competition of Fe(III) with Cd for binding sites of fulvic acids. The effect of Fe might be overestimated, because DGT measurements showed that a high proportion of dissolved Fe is present in a non-labile (inert) form, which applies for colloidal iron oxides.

DGT-labile Mn concentrations were similar to dissolved concentrations, indicating that Mn is present as dissolved Mn(II) in the water, with major species being free ions and inorganic complexes as predicted by the model. Nevertheless DGT-labile concentrations were somewhat higher than dissolved concentrations. Oxidation and precipitation of Mn(II) during sampling and filtration of water samples may lead to an underestimation of dissolved Mn concentrations in water. Such anomalous results were also obtained in other studies [26, 119].

Compared to Cd a higher proportion of Cu (12-43%) is present as non-labile species, according to the DGT measurements. Modelled labile Cu concentrations were lower than DGT-labile concentrations. It is possible that Cu is additionally bound to small organic ligands which were measured with the DGT technique, but were not included in the model. Such organic ligands with lower molecular weights than fulvic acids were identified with LC-OCD. On the other hand concentrations of calculated labile Cu-fulvic acid complexes might be underestimated, considering that diffusion coefficients of metal complexes with organic matter in natural freshwaters may be larger than diffusion coefficients of metal complexes with extracted fulvic acids under laboratory conditions [120]. Percentages of DGT-labile Cu and Zn in our study were comparable to measurements in another freshwater stream [26, 121].

The DGT technique is a useful tool to determine labile metal concentrations in water, because it measures average concentrations over time. However, long deployment times were needed in our study under conditions of low dissolved Cd concentrations. In contrast changes of dissolved cadmium concentrations were small and short, which made it difficult to measure changes of DGT-labile cadmium concentrations. Under such dynamic conditions speciation techniques with higher temporal resolution would be more appropriate, for instance gel integrated microelectrode in combination with a voltammetric in situ profiling systems (VIP-GIME) [122]. An exclusive feature of this speciation technique is that it allows the simultaneous measurement of total metal concentrations as well as concentrations of dynamic species and free metal ions.

4.5.2. Metal accumulation in periphyton

The extent of Cd accumulation in periphyton depends on the concentration and speciation of Cd and of other metals, which can compete for Cd uptake. Moreover chelation, sequestration and release of Cd influence its uptake and residence time within the cell [87].

Periphyton responded sensitively to the dynamic changes of total dissolved cadmium concentrations in water, from which a large fraction was present in labile form (Figure 4.1 a and 4.2 a). The calculated average labile Cd concentrations were comparable to the measured DGT-labile concentrations, being constant during the study. However, calculated labile Cd concentrations at discrete time points of the experiment, revealed an increase by a factor of 1.4 and 2 during the first and second rain event respectively, and concentrations were probably even higher at maximum concentrations of dissolved Cd where no modelling was performed. From the modelling results two possibilities arise, which Cd species could control bioaccumulation of Cd in this study.

Considering only fulvic acids in the model, the accumulation might depend on the free Cd ion concentrations, which were the dominant metal species. Several Cd uptake experiments with algae in defined media showed a dependence of Cd accumulation on the free metal ion concentration, which were as low as in our study [74, 97, 107]. Alternatively, if free Cd concentrations were overestimated in the model it might be possible that Cd accumulation depends on the labile inorganic and/or organic Cd complexes. In the case of low free metal ion concentrations in the bulk solution, the uptake fluxes of Cd by algae may result in diffusion limitation of free Cd ions in the diffusion layer, making the labile form the bioavailable species. This dependence was shown in a theoretical study [82] and was also demonstrated for the accumulation of Cu in periphyton at low free Cu concentrations [86]. Dependence of Cd accumulation on labile Cd was demonstrated in a previous study, in which Cd speciation was modified under semi-controlled conditions by adding NTA as a ligand [123].

The different courses of Cd accumulation in periphyton during the rain events might also depend on the Zn and Mn concentrations in water. Competition experiments with a marine diatom [106] indicate that Cd, Zn, and Mn compete with one another for binding to the Mn uptake system and the extent of Cd uptake depends on the ratios of $Cd^{2+} : Zn^{2+}$ and $Cd^{2+}: Mn^{2+}$. Considering that Cd uptake was observed in presence of Zn and Mn excess, indicates a very high affinity of Cd to the Mn transport sites. Periphyton accumulated Cd despite an excess of Zn and Mn in the present and in a previous study [13] and it is thus likely that the dynamic variations of Zn and Mn concentrations in water influenced the Cd content in periphyton.

The different courses of Cd content in periphyton after the rain events might be influenced by various mechanisms regulating the intracellular Cd content. Cd can be bound to intracellular ligands, for example phytochelatins (PC). These metal-binding peptides are strongly induced in algae by Cd even at low concentrations [89, 90] and are involved in Cd detoxification. Cd-PC complexes might then be sequestered in vacuoles [88] or released from cells [18]. Intracellular binding of Cd to PC might be further influenced by the repeated exposure of periphyton to elevated metal concentrations during the rain events, which also differed in length and intensity.

4.5.3. Environmental implications

This study has shown that periphyton accumulates Cd in response to small variations of low dissolved Cd concentrations in water and in the presence of an excess of Zn and Mn. It can be expected that Cd accumulation in periphyton will be higher at higher Cd and/or lower Mn and Zn concentrations in water. Higher Cd concentrations in water are also expected to increase the bioavailable fraction. Since the release of Cd from periphyton is slow, it is possible that repeated exposures to high dissolved Cd concentrations will cause a continuous increase of Cd content in periphyton. Since periphyton is the most important primary producer in aquatic systems, it is expected that elevated Cd contents will not only affect periphyton, but also other trophic levels. This might occur by selection of metal-tolerant species and biomagnification.

Outlook

Outlook

In this work artificial channels have been shown to be a useful tool to examine the relationships between Cd speciation and accumulation in periphyton under natural freshwater conditions. It has been shown that periphyton responds sensitively to environmentally relevant Cd concentrations and that labile metal species can control bioaccumulation. The relationships between Cd speciation and accumulation obtained in the field study were less clear. Several important questions remained unanswered in this work.

The kinetics of Cd uptake in periphyton were slower when compared to the kinetics of single algal species in chemically defined media. It was hypothesized that the matrix of periphyton might limit the diffusion of metals to algae, but the capacity of the matrix in binding metals was not examined. Despite that matrix components contain high amounts of negatively charged functional groups (carboxylic, phosphate, sulphate) [124], a study with microbial biofilms has shown that only low amounts of Cd were found in the matrix [125]. Studies on the metal complexing capacity of periphyton matrix components are needed to elucidate their effects on metal bioavailability. Several methods for the separation of the matrix from algae, bacteria or sewage sludge are reported. They include different physical, chemical and mechanical treatments [126]. Characterization of the matrix could be performed with size-exclusion chromatography in connection to on-line, high sensitivity organic carbon detection (LC-OCD) [32-34]. Sorption experiments could be carried out in defined media with constant Cd and different matrix concentrations with simultaneous measurements of free Cd concentrations. In subsequent experiments, the effect of the matrix on Cd uptake could be investigated by exposing periphyton at constant Cd and increasing matrix concentrations. A more direct approach could be to perform Cd uptake experiments with intact periphyton biofilms and algae, from which the matrix has been separated.

All three studies of this work have shown that Cd is accumulated by periphyton despite an excess of zinc and manganese. Cd uptake in algae is reported to occur via Zn [107] or Mn transporters [97, 106] or a transporter acting in the transport of Mn,

Outlook

Fe and Cu [108]. Competitive interactions of other metals on Cd uptake in periphyton could be performed in chemically controlled media, where speciation could be varied and calculated precisely. The approach would be to expose periphyton to Cd at a constant free ion concentration together with varying free concentrations of another metal.

Periphyton accumulated substantial amounts of Cd in the channel experiments and responded sensitively to changes of Cd concentrations in water in the field study. Whether such intracellular concentrations produce toxicity to periphytic algae was not investigated. Endpoints which could be used for Cd toxicity are inhibition of photosynthesis and production of reactive oxygen species (ROS). Inhibition of photosynthesis can be determined by measuring the photosynthetic yield using the pulse amplitude modulation chlorophyll fluorescence technique (PhytoPAM). The production of ROS can be determined with a fluorescence dye [127]. The exposure of periphyton to increasing Cd concentrations would give information about dose-response relationships. Cd concentrations inhibiting photosynthesis could be then used for pulse exposure experiments, to test whether periphyton is capable to recover from this stress or not.

The technique of diffusion gradients in thin-films (DGT) with non-restricted gels allowed the simultaneous measurement of different Cd species. The measurement was easier in the channels than in the field study, where the resolution time of DGTs was longer compared to the changes of metal concentrations in water. In order to relate accumulation data to a particular metal species, DGTs with restricted and non-restricted gels could be combined [40] together with a technique for measuring free metal ions. Moreover, since each speciation technique has its own required analysis time, each of them is suitable either for systems with stable or dynamic conditions.

References

References

[1] Rand GM. 1995. Fundamentals of aquatic toxicology: Effects, environmental fate, and risk assessment, Second edition ed. Taylor & Francis, Washington, D.C.

[2] Morel FMM, Hudson RJM, Price NM. 1991. Limitation of Productivity by Trace-Metals in the Sea. *Limnology and Oceanography* 36: 1742-1755.

[3] Morin S, Duong TT, Dabrin A, Coynel A, Herlory O, Baudrimont M, Delmas F, Durrieu G, Schafer J, Winterton P, Blanc G, Coste M. 2008. Long-term survey of heavy-metal pollution, biofilm contamination and diatom community structure in the Riou Mort watershed, South-West France. *Environmental Pollution* 151: 532-542.

[4] Lehmann V, Tubbing GMJ, Admiraal W. 1999. Induced metal tolerance in microbenthic communities from three lowland rivers with different metal loads. *Archives of Environmental Contamination and Toxicology* 36: 384-391.

[5] Admiraal W, Blanck H, Buckert-De Jong M, Guasch H, Ivorra N, Lehmann V, Nystrom BAH, Paulsson M, Sabater S. 1999. Short-term toxicity of zinc to microbenthic algae and bacteria in a metal polluted stream. *Water Research* 33: 1989-1996.

[6] Lewis MA, Quarles RL, Dantin DD, Moore JC. 2004. Evaluation of a Florida coastal golf complex as a local and watershed source of bioavailable contaminants. *Marine Pollution Bulletin* 48: 254-262.

[7] Tessier A, Turner DR. 1995. Metal speciation and bioavailability in aquatic systems. Wiley, Chichester, West Sussex, England.

[8] Mason AZ, Jenkins KD. 1995. Metal detoxification in aquatic organisms. In Tessier A, Turner DR, eds, *Metal speciation and bioavailability in aquatic systems*. John Wiley, New York, USA, pp 479-608.

[9] Campbell PGC. 2006. Cadmium - A priority pollutant. *Environmental Chemistry* 3:387-388.

[10] Dobson S. 1992. Cadmium - environmental aspects. World Health Organization, Geneva, Switzerland.

[11] Xue HB, Sigg L. 1998. Cadmium speciation and complexation by natural organic ligands in fresh water. *Analytica Chimica Acta* 363:249-259.

[12] Sigg L, Black F, Buffle J, Cao J, Cleven R, Davison W, Galceran J, Gunkel P, Kalis E, Kistler D, Martin M, Noel S, Nur Y, Odzak N, Puy J, Van Riemsdijk W, Temminghoff E, Tercier-Waeber ML, Toepperwien S, Town RM, Unsworth E, Warnken KW, Weng LP, Xue HB, Zhang H. 2006. Comparison of analytical techniques for dynamic trace metal speciation in natural freshwaters. *Environmental Science and Technology* 40:1934-1941.

References

[13] Bradac P, Navarro E, Odzak N, Behra R, Sigg L. 2009. Kinetics of cadmium accumulation in periphyton under freshwater conditions. *Environmental Toxicology and Chemistry* 28 (10): 2108–2116.

[14] Morel FMM, Hering JG. 1993. Principles and applications of aquatic chemistry. Wiley, New York, USA.

[15] Sigg L, Stumm W. 1989. Aquatische Chemie: Eine Einführung in die Chemie wässriger Lösungen und natürlicher Gewässer. vdf, Hochschulverlag ETH Zürich, Teubner, Zürich, Stuttgart.

[16] Sigg L, Behra R. 2005. Speciation and bioavailability of trace metals in freshwater environments. In Sigel A, Sigel H, Sigel RKO, eds, *Biogeochemistry, Availability, and Transport of Metals in the Environment Metal ions in biological systems*. Vol 44. Taylor and Francis Group, Boca Raton, Fl, USA, pp 47-73.

[17] Tipping E. 2002. Cation Binding by Humic Substances. Cambridge University Press, Cambridge.

[18] Lee JG, Ahner BA, Morel FMM. 1996. Export of cadmium and phytochelatin by the marine diatom Thalassiosira weissflogii. *Environmental Science and Technology* 30:1814-1821.

[19] Wei LP, Ahner BA. 2005. Sources and sinks of dissolved phytochelatin in natural seawater. *Limnology and Oceanography* 50: 13-22.

[20] Giger W, Schaffner C, Kari FG, Ponusz H, Reichert P, Wanner O. 1991. *EAWAG news*. pp 27-31.

[21] Buffle J. 1988. Complexation reactions in aquatic systems: an analytical approach. John Wiley and Sons Inc., Chichester.

[22] Xue H-B, Sigg L. 2002. A Review of Competitive Ligand-Exchange-Voltammetric Methods for Speciation of Trace Metals in Freshwater. In Rozan TF, Taillefert M, eds, *Environmental Electrochemistry: Analyses of Trace Element Biogeochemistry*. American Chemical Society, Washington, pp 336-370.

[23] van Leeuwen HP, Town RM, Buffle J, Cleven R, Davison W, Puy J, van Riemsdijk WH, Sigg L. 2005. Dynamic speciation analysis and bioavailability of metals in aquatic systems. *Environmental Science and Technology* 39: 8545-8556.

[24] Scally S, Davison W, Zhang H. 2003. In situ measurements of dissociation kinetics and labilities of metal complexes in solution using DGT. *Environmental Science and Technology* 37:1379-1384.

[25] Van Leeuwen HP. 2000. Dynamic aspects of in situ speciation processes and techniques. In Buffle J, Horvai G, eds, *In situ monitoring of aquatic systems chemical analysis and speciation*. John Wiley and Sons, Chichester, p. 623.

References

[26] Unsworth ER, Warnken KW, Zhang H, Davison W, Black F, Buffle J, Cao J, Cleven R, Galceran J, Gunkel P, Kalis E, Kistler D, Van Leeuwen HP, Martin M, Noel S, Nur Y, Odzak N, Puy J, Van Riemsdijk W, Sigg L, Temminghoff E, Tercier-Waeber ML, Toepperwien S, Town RM, Weng LP, Xue HB. 2006. Model predictions of metal speciation in freshwaters compared to measurements by in situ techniques. *Environmental Science and Technology* 40:1942-1949.

[27] Tipping E. 1998. Humic ion-binding model VI: An improved description of the interactions of protons and metal ions with humic substances. *Aquatic Geochemistry* 4: 3-48.

[28] Tipping E, Hurley MA. 1992. A Unifying Model of Cation Binding by Humic Substances. *Geochimica et Cosmochimica Acta* 56: 3627-3641.

[29] Gustafsson JP. 2006. Visual MINTEQ. 2.51 ed. Royal Institute of Technology, Department of Land and Water Resources Engineering, Stockholm, Sweden.

[30] Benedetti MF, Milne CJ, Kinniburgh DG, VanRiemsdijk WH, Koopal LK. 1995. Metal-Ion Binding to Humic Substances - Application of the Nonideal Competitive Adsorption Model. *Environmental Science and Technology* 29: 446-457.

[31] Gustafsson JP. 2001. Modeling the acid-base properties and metal complexation of humic substances with the Stockholm Humic Model. *Journal of Colloid and Interface Science* 244: 102-112.

[32] Huber SA, Gluschke M. 1998. Chromatographic characterization of TOC in process water treatment. *Ultrapure Water*: 48-52.

[33] Huber S, Frimmel FH. 1992. A Liquid-Chromatographic System with Multi-Detection for the Direct Analysis of Hydrophilic Organic-Compounds in Natural-Waters. *Fresenius Journal of Analytical Chemistry* 342: 198-200.

[34] Huber SA, Frimmel FH. 1996. Size-exclusion chromatography with organic carbon detection (LC-OCD): A fast and reliable method for the characterization of hydrophilic organic matter in natural waters. *Vom Wasser*. Vol 86. VCH, Weinheim, Germany, pp 277-290.

[35] Tomaszewski L, Buffle J, Galceran J. 2003. Theoretical and analytical characterization of a flow-through permeation liquid membrane with controlled flux for metal speciation measurements. *Analytical Chemistry* 75: 893-900.

[36] Van Leeuwen HP, Town RM. 2003. Electrochemical metal speciation analysis of chemically heterogeneous samples: The outstanding features of stripping chronopotentiometry at scanned deposition potential. *Environmental Science and Technology* 37: 3945-3952.

References

[37] Town RM, van Leeuwen HP. 2004. Depletive stripping chronopotentiometry: A major step forward in electrochemical stripping techniques for metal ion speciation analysis. *Electroanalysis* 16: 458-471.

[38] Davison W, Zhang H. 1994. In-situ speciation measurements of trace components in natural-waters using thin-film gels. *Nature* 367: 546-548.

[39] Zhang H, Davison W. 1995. Performance-Characteristics of Diffusion Gradients in Thin-Films for the in-Situ Measurement of Trace-Metals in Aqueous-Solution. *Analytical Chemistry* 67: 3391-3400.

[40] Zhang H, Davison W. 2000. Direct in situ measurements of labile inorganic and organically bound metal species in synthetic solutions and natural waters using diffusive gradients in thin films. *Analytical Chemistry* 72: 4447-4457.

[41] Tercier ML, Parthasarathy N, Buffle J. 1995. Reproducible, reliable and rugged Hg-plated IR-based microelectrode for in-situ measurements in natural-waters. *Electroanalysis* 7:55-63.

[42] Tercier-Waeber M, Confalonieri F, Riccardi G, Sina A, Graziottin F, Buffle J. 2003. Multi physical-chemical profiler for real-time automated in situ monitoring of specific fractions of trace metals and master variables. 12th International Conference on Heavy Metals in the Environment, Grenoble, France, May 26-30, pp 1297-1300.

[43] Temminghoff EJM, Plette ACC, Van Eck R, Van Riemsdijk WH. 2000. Determination of the chemical speciation of trace metals in aqueous systems by the Wageningen Donnan Membrane Technique. *Analytica Chimica Acta* 417: 149-157.

[44] Weng LP, Van Riemsdijk WH, Temminghoff EJM. 2005. Kinetic aspects of donnan membrane technique for measuring free trace cation concentration. *Analytica Chimica Acta* 77: 2852-2861.

[45] Parthasarathy N, Pelletier M, Buffle J. 1997. Hollow fiber based supported liquid membrane: a novel analytical system for trace metal analysis. *Analytica Chimica Acta* 350: 183-195.

[46] Eriksen RS, Mackey DJ, van Dam R, Nowak B. 2001. Copper speciation and toxicity in Macquarie Harbour, Tasmania: an investigation using a copper ion selective electrode. *Marine Chemistry* 74: 99-113.

[47] Stevenson RJ, Bothwell ML, Lowe RL. 1996. Algal ecology: Freshwater benthic ecosystems. Academic Press, San Diego, USA.

[48] Flemming HC, Wingender J. 2001. Relevance of microbial extracellular polymeric substances (EPSs) - Part I: Structural and ecological aspects. *Water Science and Technology* 43: 1-8.

References

[49] Smith DJ, Underwood GJC. 2000. The production of extracellular carbohydrates by estuarine benthic diatoms: The effects of growth phase and light and dark treatment. *Journal of Phycology* 36: 321-333.

[50] Wingender J, Neu T, Flemming H-C. 1999. What are bacterial extracellular polymer substances? In Wingender J, Neu T, Flemming H-C, eds, *Microbial extracellular polymeric substances: Characterization, structure and function*, 1st ed. Springer, Heidelberg, Berlin, pp 1-19.

[51] Holding KL, Gill RA, Carter J. 2003. The relationship between epilithic periphyton (biofilm) bound metals and metals bound to sediments in freshwater systems. *Environmental Geochemistry and Health* 25: 87-93.

[52] Lombardi AT, Vieira AAH. 1998. Copper and lead complexation by high molecular weight compounds produced by Synura sp. (Chrysophyceae). *Phycologia* 37:34-39.

[53] Wuertz S, Spaeth R, Hinderberger A, Grieba T, Flemming HC, Wilderer PA. 2001. A new method for extraction of extracellular polymeric substances from biofilms and activated sludge suitable for direct quantification of sorbed metals. *Water Science and Technology* 43: 25-31.

[54] Pistocchi R, Mormile MA, Guerrini F, Isani G, Boni L. 2000. Increased production of extra- and intracellular metal-ligands in phytoplankton exposed to copper and cadmium. *Journal of Applied Phycology* 12: 469-477.

[55] Pusch M, Fiebig D, Brettar I, Eisenmann H, Ellis BK, Kaplan LA, Lock MA, Naegeli MW, Traunspurger W. 1998. The role of micro-organisms in the ecological connectivity of running waters. *Freshwater Biology* 40: 453-495.

[56] Kaplan LA, Bott TL. 1983. Microbial Heterotrophic Utilization of Dissolved Organic-Matter in a Piedmont Stream. *Freshwater Biology* 13: 363-377.

[57] Tate CM, Broshears RE, McKnight DM. 1995. Phosphate Dynamics in an Acidic Mountain Stream - Interactions Involving Algal Uptake, Sorption by Iron-Oxide, and Photoreduction. *Limnology and Oceanography* 40: 938-946.

[58] Allan JD. 1995. Stream ecology : structure and function of running waters. Chapman & Hall, London.

[59] Duong TT, Feurtet-Mazel A, Coste M, Dang DK, Boudou A. 2007. Dynamics of diatom colonization process in some rivers influenced by urban pollution (Hanoi, Vietnam). *Ecological Indicators* 7: 839-851.

[60] Ivorra N, Hettelaar J, Kraak MHS, Sabater S, Admiraal W. 2002. Responses of biofilms to combined nutrient and metal exposure. *Environmental Toxicology and Chemistry* 21: 626-632.

[61] Morin S, Duong TT, Herlory O, Feurtet-Mazel A, Coste M. 2008. Cadmium toxicity and bioaccumulation in freshwater biofilms. *Archives of Environmental Contamination and Toxicology* 54: 173-186.

References

[62] Gold C, Feurtet-Mazel A, Coste M, Boudou A. 2003. Effects of cadmium stress on periphytic diatom communities in indoor artificial streams. *Freshwater Biology* 48: 316-328.

[63] Croteau MN, Luoma SN, Stewart AR. 2005. Trophic transfer of metals along freshwater food webs: Evidence of cadmium biomagnification in nature. *Limnology and Oceanography* 50: 1511-1519.

[64] Campbell PGC, Errecalde O, Fortin C, Hiriart-Baer WR, Vigneault B. 2002. Metal bioavailability to phytoplankton - applicability of the biotic ligand model. *Comparative Biochemistry and Physiology C-Toxicology and Pharmacology* 133: 189-206.

[65] Simkiss K, Taylor MG. 1995. Transport of metals across membranes. In Tessier A, Turner DR, eds, *Metal speciation and bioavailability in aquatic systems*, John Wiley and Sons ed. Vol 3-IUPAC Series. John Wiley and Sons, Chichester, pp 1-44.

[66] Sunda WG, Guillard RRL. 1976. Relationship between cupric ion activity and toxicity of copper to phytoplankton. *Journal of Marine Research* 34: 511-529.

[67] Anderson MA, Morel FMM, Guillard RRL. 1978. Growth Limitation of a Coastal Diatom by Low Zinc Ion Activity. *Nature* 276: 70-71.

[68] Allen HE, Hall RH, Brisbin TD. 1980. Metal Speciation - Effects on Aquatic Toxicity. *Environmental Science and Technology* 14: 441-443.

[69] Bates SS, Tessier A, Campbell PGC, Buffle J. 1982. Zinc Adsorption and Transport by Chlamydomonas-Variabilis and Scenedesmus-Subspicatus (Chlorophyceae) Grown in Semi-Continuous Culture. *Journal of Phycology* 18: 521-529.

[70] Campbell PGC. 1995. Interactions between trace metals and aquatic organisms: A critique of the free-ion activity model. In Tessier A, Turner DR, eds, *Metal speciation and bioavailability in aquatic systems*. Vol 3-IUPAC Series. John Wiley and Sons, Chichester, pp 45-102.

[71] Sunda WG, Huntsman SA. 1992. Feedback Interactions between Zinc and Phytoplankton in Seawater. *Limnology and Oceanography* 37: 25-40.

[72] Sunda WG, Huntsman SA. 1998. Control of Cd concentrations in a coastal diatom by interactions among free ionic Cd, Zn, and Mn in seawater. *Environmental Science and Technology* 32: 2961-2968.

[73] Knauer K, Behra R, Sigg L. 1997. Effects of free Cu^{2+} and Zn^{2+} ions on growth and metal accumulation in freshwater algae. *Environmental Toxicology and Chemistry* 16: 220-229.

[74] Kola H, Wilkinson KJ. 2005. Cadmium uptake by a green alga can be predicted by equilibrium modelling. *Environmental Science and Technology* 39: 3040-3047.

References

[75] Fortin C, Campbell PGC. 2001. Thiosulfate enhances silver uptake by a green alga: Role of anion transporters in metal uptake. *Environmental Science and Technology* 35: 2214-2218.

[76] Errecalde O, Seidl M, Campbell PGC. 1998. Influence of a low molecular weight metabolite (citrate) on the toxicity of cadmium and zinc to the unicellular green alga Selenastrum capricornutum: An exception to the free-ion model. *Water Research* 32: 419-429.

[77] Errecalde O, Campbell PGC. 2000. Cadmium and zinc bioavailability to Selenastrum capricornutum (Chlorophyceae): accidental metal uptake and toxicity in the presence of citrate. *Journal of Phycology* 36: 473-483.

[78] Phinney JT, Bruland KW. 1994. Uptake of Lipophilic Organic Cu, Cd, and Pb Complexes in the Coastal Diatom Thalassiosira-Weissflogii. *Environmental Science and Technology* 28: 1781-1790.

[79] Croot PL, Karlson B, Van Elteren JT, Kroon JJ. 1999. Uptake of Cu-64-oxine by marine phytoplankton. *Environmental Science and Technology* 33: 3615-3621.

[80] Turner A, Mawji E. 2005. Octanol-solubility of dissolved and particulate trace metals in contaminated rivers: implications for metal reactivity and availability. *Environmental Pollution* 135: 235-244.

[81] Di Toro DM, Allen HE, Bergman HL, Meyer JS, Paquin PR, Santore RC. 2001. Biotic ligand model of the acute toxicity of metals. 1. Technical basis. *Environmental Toxicology and Chemistry* 20: 2383-2396.

[82] Van Leeuwen HP. 1999. Metal speciation dynamics and bioavailability: Inert and labile complexes. *Environmental Science and Technology* 33: 3743-3748.

[83] Slaveykova VI, Wilkinson KJ. 2005. Predicting the bioavailability of metals and metal complexes: Critical review of the biotic ligand model. *Environmental Chemistry* 2: 9-24.

[84] Hudson RJM. 1998. Which aqueous species control the rates of trace metal uptake by aquatic biota? Observations and predictions of non-equilibrium effects. *Science of the Total Environment* 219: 95-115.

[85] Fortin C, Campbell PGC. 2000. Silver uptake by the green alga Chlamydomonas reinhardtii in relation to chemical speciation: Influence of chloride. *Environmental Toxicology and Chemistry* 19: 2769-2778.

[86] Meylan S, Behra R, Sigg L. 2003. Accumulation of copper and zinc in periphyton in response to dynamic variations of metal speciation in freshwater. *Environmental Science and Technology* 37: 5204-5212.

[87] Clemens S. 2001. Molecular mechanisms of plant metal tolerance and homeostasis. *Planta* 212:475-486.

References

[88] Vogelilange R, Wagner GJ. 1990. Subcellular-localization of cadmium and cadmium-binding peptides in tobacco-leaves - Implication of a transport function for cadmium-binding peptides. *Plant Physiology* 92: 1086-1093.

[89] Ahner BA, Price NM, Morel FMM. 1994. Phytochelatin Production by Marine-Phytoplankton at Low Free Metal-Ion Concentrations - Laboratory Studies and Field Data from Massachusetts Bay. *Proceedings of the National Academy of Sciences of the United States of America* 91: 8433-8436.

[90] Le Faucheur S, Behra R, Sigg L. 2005. Phytochelatin induction, cadmium accumulation, and algal sensitivity to free cadmium ion in Scenedesmus vacuolatus. *Environmental Toxicology and Chemistry* 24: 1731-1737.

[91] Ortiz DF, Kreppel L, Speiser DM, Scheel G, McDonald G, Ow DW. 1992. Heavy-metal tolerance in the fission yeast requires an ATP-binding cassette-type vacuolar membrane transporter. *Embo Journal* 11:3491-3499.

[92] Stephenson M, Turner MA. 1993. A Field-Study of Cadmium Dynamics in Periphyton and in Hyalella-Azteca (Crustacea, Amphipoda). *Water Air and Soil Pollution* 68: 341-361.

[93] Hill WR, Larsen IL. 2005. Growth dilution of metals in microalgal biofilms. *Environmental Science and Technology* 39: 1513-1518.

[94] Wang WX, Rainbow PS. 2006. Subcellular partitioning and the prediction of cadmium toxicity to aquatic organisms. *Environmental Chemistry* 3:395-399.

[95] Wang MJ, Wang WX. 2008. Cadmium toxicity in a marine diatom as predicted by the cellular metal sensitive fraction. *Environmental Science and Technology* 42: 940-946.

[96] Miao AJ, Wang WX, Juneau P. 2005. Comparison of Cd, Cu, and Zn toxic effects on four marine phytoplankton by pulse-amplitude-modulated fluorometry. *Environmental Toxicology and Chemistry* 24: 2603-2611.

[97] Sunda WG, Huntsman SA. 2000. Effect of Zn, Mn, and Fe on Cd accumulation in phytoplankton: Implications for oceanic Cd cycling. *Limnology and Oceanography* 45: 1501-1516.

[98] Vigneault B, Campbell PGC. 2005. Uptake of cadmium by freshwater green algae: Effects of pH and aquatic humic substances. *Journal of Phycology* 41: 55-61.

[99] Brooks BW, Stanley JK, White JC, Turner PK, Wu KB, La Point TW. 2004. Laboratory and field responses to cadmium: An experimental study in effluent-dominated stream mesocosms. *Environmental Toxicology and Chemistry* 23: 1057-1064.

[100] Selby DA, Ihnat JM, Messer JJ. 1985. Effects of Subacute Cadmium Exposure on a Hardwater Mountain Stream Microcosm. *Water Research* 19: 645-655.

References

[101] Hill WR, Bednarek AT, Larsen IL. 2000. Cadmium sorption and toxicity in autotrophic biofilms. *Canadian Journal of Fisheries and Aquatic Sciences* 57: 530-537.

[102] Zhang H. 1997. Practical Guide for Making Gels for Dgt and Det and Practical Guide for Using Dgt and Det. DGT Research Ltd., Lancaster, UK.

[103] Navarro E, Robinson CT, Wagner B, Behra R. 2007. Influence of ultraviolet radiation on UVR-absorbing compounds in freshwater algal biofilms and Scenedesmus vacuolatus cultures. *Journal of Toxicology and Environmental Health - Part A* 70: 760-767.

[104] Murray AP, Gibbs CF, Longmore AR, Flett DJ. 1986. Determination of Chlorophyll in Marine Waters - Intercomparison of a Rapid Hplc Method with Full Hplc, Spectrophotometric and Fluorometric Methods. *Marine Chemistry* 19: 211-227.

[105] McGeer JC, Brix KV, Skeaff JM, DeForest DK, Brigham SI, Adams WJ, Green A. 2003. Inverse relationship between bioconcentration factor and exposure concentration for metals: Implications for hazard assessment of metals in the aquatic environment. *Environmental Toxicology and Chemistry* 22: 1017-1037.

[106] Sunda WG, Huntsman SA. 1996. Antagonisms between cadmium and zinc toxicity and manganese limitation in a coastal diatom. *Limnology and Oceanography* 41: 373-387.

[107] Topperwien S, Behra R, Sigg L. 2007. Competition among zinc, manganese, and cadmium uptake in the freshwater alga Scenedesmus vacuolatus. *Environmental Toxicology and Chemistry* 26: 483-490.

[108] Rosakis A, Koster W. 2005. Divalent metal transport in the green microalga Chlamydomonas reinhardtii is mediated by a protein similar to prokaryotic Nramp homologues. *Biometals* 18: 107-120.

[109] Morelli E, Scarano G. 2001. Synthesis and stability of phytochelatins induced by cadmium and lead in the marine diatom Phaeodactylum tricornutum. *Marine Environmental Research* 52: 383-395.

[110] Newman MC. 1989. Appropriateness of aufwuchs as a monitor of bioaccumulation. *Environmental Pollution* 60: 83-100.

[111] Unsworth ER, Zhang H, Davison W. 2005. Use of diffusive gradients in thin films to measure cadmium speciation in solutions with synthetic and natural ligands: Comparison with model predictions. *Environmental Science and Technology* 39: 624-630.

[112] Scally S, Zhang H, Davison W. 2004. Measurements of lead complexation with organic ligands using DGT. *Australian Journal of Chemistry* 57: 925-930.

References

[113] Gesellschaft W. 1986. Bestimmung des Chlorophyll-a-Gehaltes von Oberflächenwasser. *Deutsche Einheitsverfahren zur Wasser-, Abwasser- und Schlammuntersuchung*. Wiley-VCH, Weinheim, Germany.

[114] Frimmel FH, Grenz R, Kordik E, Dietz F. 1989. Nitrotriacetate (NTA) and ethylenedinitrilotetraacetate (EDTA) in rivers of the Federal Republic of Germany.

[115] Horiet J-P. 1990. Development of the concentration of the laundry detergent phosphate substitute "NTA" in Switzerland waters, situation 1990. *BUWAL-Bulletin*. Vol 3/90.

[116] Malaiyandi M, Williams DT, Ogrady R. 1979. National survey of nitrilotriacetic acid in Canadian drinking-water. *Environmental Science and Technology* 13: 59-62.

[117] Behra R, Genoni GP, Sigg L. 1993. Water quality criteria for metals and metalloids in running water: scientific basis. *Gas, Wasser, Abwasser* 73:942-951.

[118] Denney S, Sherwood J, Leyden J. 1999. In situ measurements of labile Cu, Cd and Mn in river waters using DGT. *Science of the Total Environment* 239: 71-80.

[119] Odzak N, Kistler D, Xue HB, Sigg L. 2002. In situ trace metal speciation in a eutrophic lake using the technique of diffusion gradients in thin films (DGT). *Aquatic Sciences* 64: 292-299.

[120] Warnken KW, Davison W, Zhang H. 2008. Interpretation of in situ speciation measurements of inorganic and organically complexed trace metals in freshwater by DGT. *Environmental Science and Technology* 42: 6903-6909.

[121] Meylan S, Odzak N, Behra R, Sigg L. 2004. Speciation of copper and zinc in natural freshwater: comparison of voltammetric measurements, diffusive gradients in thin films (DGT) and chemical equilibrium models. *Analytica Chimica Acta* 510: 91-100.

[122] Tercier-Waeber ML, Confalonieri F, Riccardi G, Sina A, Noel S, Buffle J, Graziottin F. 2005. Multi Physical-Chemical profiler for real-time in situ monitoring of trace metal speciation and master variables: Development, validation and field applications. *Marine Chemistry* 97: 216-235.

[123] Bradac P, Behra R, Sigg L. 2009. Accumulation of cadmium in periphyton under various freshwater speciation conditions. *Environmental Science and Technology* 43: 7291-7296.

[124] Sutherland IW. 1984. Microbial exopolysaccharides - their role in microbial adhesion in aqueous systems. *Critical Reviews in Microbiology* 10: 173-201.

[125] Wuertz S, Spaeth R, Hinderberger A, Grieba T, Flemming HC, Wilderer PA. 2000. A new method for extraction of extracellular polymeric substances from biofilms and activated sludge suitable for direct quantification of sorbed metals. 1st IWA International Conference on Microbial Extracellular Polymeric Substances, Mulheim, Germany, Sep 18-20, pp 25-31.

[126] Azeredo J, Lazarova V, Oliveira R. 1999. Methods to extract the exopolymeric matrix from biofilms: A comparative study. *Water Science and Technology* 39: 243-250.

[127] Szivák I, Behra R, Sigg L. 2009. Metal induced reactive oxygen species production in Chlamydomonas reinhardtii (Chlorophyceae). *Journal of Phycology* 45(2): 427-435.

Appendix

Appendix

Supporting information to Chapter 4

Modelled metal species (S1)

Modelled metal species assuming Fe being present as dissolved Fe(III) (+Fe(III)) or precipitated as Fe(III) oxides (-Fe(III)).

		Time [days]	-0.3	0.9	1.2	1.7	1.9	2.7
Cd		Dissolved	0.10	0.17	0.16	0.21	0.17	0.10
(-Fe(III))		Free	0.071	0.101	0.102	0.140	0.096	0.058
		% free	68	60	65	66	56	61
		Inorganic complexes	0.026	0.056	0.040	0.060	0.060	0.028
		% inorganic complexes	25	33	26	28	35	30
		FA-complexes	0.007	0.012	0.015	0.012	0.015	0.009
		% FA-complexes	7	7	9	6	9	10
		Calculated labile	0.10	0.16	0.14	0.20	0.16	0.09
		% calculated labile	94	94	93	95	93	92
Cd		Free	0.074	0.106	0.109	0.146	0.102	0.063
(+Fe(III))		% free	71	63	70	69	60	66
		Inorganic complexes	0.028	0.059	0.043	0.062	0.065	0.031
		% inorganic complexes	27	35	28	29	38	32
		FA-complexes	0.002	0.004	0.004	0.004	0.004	0.002
		% FA-complexes	2.4	2.3	2.5	2.0	2.5	2.3
		Calculated labile	0.10	0.17	0.15	0.21	0.17	0.09
		% calculated labile	98	98	98	98	98	98
Zn		Dissolved	636	502	235	843	214	197
(-Fe(III))		Free	362	229	137	493	86	92
		% free	57	46	58	58	40	47
		Inorganic complexes	231	237	76	288	110	88
		% inorganic complexes	36	47	33	34	51	45
		FA-complexes	44	36	21	62	19	17
		% FA-complexes	7	7	9	7	9	9
		Calculated labile	602	473	217	793	199	183
		% calculated labile	95	94	93	94	93	93
Zn		Free	376	239	145	513	90	97
(+Fe(III))		% free	59	48	62	61	42	50
		Inorganic complexes	240	246	81	300	116	93
		% inorganic complexes	38	49	34	36	54	47
		FA-complexes	21.0	16.7	8.7	29.3	7.8	6.7
		% FA-complexes	3.3	3.3	3.7	3.5	3.7	3.4
		Calculated labile	620	488	228	820	208	191
		% calculated labile	97	97	97	97	97	97

Appendix

	Time [days]	2.9	3.7	3.9	4.7	4.9	5.7
Cd (-Fe(III))	Dissolved	0.18	0.11	0.10	0.07	0.06	0.06
	Free	0.122	0.075	0.060	0.047	0.036	0.035
	% free	67	67	62	70	64	63
	Inorganic complexes	0.050	0.030	0.029	0.013	0.014	0.016
	% inorganic complexes	27	27	30	20	25	28
	FA-complexes	0.010	0.007	0.008	0.006	0.006	0.005
	% FA-complexes	6	6	8	10	11	9
	Calculated labile	0.17	0.11	0.09	0.06	0.05	0.05
	% calculated labile	96	95	93	92	91	93
Cd (+Fe(III))	Free	0.127	0.078	0.064	0.051	0.039	0.037
	% free	70	70	66	76	70	68
	Inorganic complexes	0.052	0.032	0.031	0.014	0.016	0.017
	% inorganic complexes	29	28	32	21	28	30
	FA-complexes	0.003	0.002	0.002	0.002	0.002	0.001
	% FA-complexes	1.8	1.9	2.4	2.8	2.9	2.3
	Calculated labile	0.18	0.11	0.10	0.07	0.06	0.05
	% calculated labile	99	98	98	98	98	98
Zn (-Fe(III))	Dissolved	449	239	171	156	107	98
	Free	246	127	79	98	52	46
	% free	55	53	46	63	49	47
	Inorganic complexes	170	94	77	41	43	44
	% inorganic complexes	38	39	45	26	40	45
	FA-complexes	32	18	15	17	11	8
	% FA-complexes	7	8	9	11	11	8
	Calculated labile	423	225	159	143	98	92
	% calculated labile	94	94	93	91	91	94
Zn (+Fe(III))	Free	257	134	84	105	56	49
	% free	57	56	49	67	52	50
	Inorganic complexes	178	98	81	44	46	46
	% inorganic complexes	40	41	47	28	43	47
	FA-complexes	13.8	7.3	6.2	6.8	4.6	3.2
	% FA-complexes	3.1	3.1	3.6	4.4	4.3	3.2
	Calculated labile	438	233	166	151	103	96
	% calculated labile	98	98	97	96	97	97

Appendix

		Time [days]	-0.3	0.9	1.2	1.7	1.9	2.7
Mn	(-Fe(III))	Dissolved	64	105	104	174	134	137
		Free	43	60	72	120	70	80
		% free	67	57	70	69	52	59
		Inorganic complexes	21	45	31	54	64	56
		% inorganic complexes	33	43	30	31	48	41
		FA-complexes	0.04	0.06	0.09	0.18	0.08	0.07
		% FA-complexes	0.06	0.06	0.08	0.10	0.06	0.05
		Calculated labile	64	105	104	174	134	137
		% calculated labile	100	100	100	100	100	100
Mn	(+Fe(III))	Free	43	60	72	120	70	80
		% free	67	57	70	69	52	59
		Inorganic complexes	21	45	31	54	64	56
		% inorganic complexes	33	43	30	31	48	41
		FA-complexes	0.041	0.057	0.084	0.170	0.074	0.071
		% FA-complexes	0.06	0.05	0.08	0.10	0.05	0.05
		Calculated labile	64	105	104	174	134	137
		% calculated labile	100	100	100	100	100	100
Cu	(-Fe(III))	Dissolved	21	33	28	71	44	23
		Free	0.013	0.014	0.013	0.063	0.012	0.007
		% free	0.061	0.041	0.047	0.088	0.026	0.029
		Inorganic complexes	0.67	1.18	0.57	2.80	1.25	0.53
		% inorganic complexes	3.1	3.6	2.0	3.9	2.8	2.3
		FA-complexes	21	32	28	69	43	23
		% FA-complexes	97	96	98	96	97	98
		Calculated labile	4.8	7.5	6.2	16.6	9.9	5.0
		% calculated labile	23	23	22	23	22	22
Cu	(+Fe(III))	Free	0.056	0.059	0.085	0.277	0.059	0.042
		% free	0.26	0.18	0.30	0.39	0.13	0.18
		Inorganic complexes	2.9	5.1	3.6	12.3	6.3	3.3
		% inorganic complexes	13	16	13	17	14	14
		FA-complexes	18	28	25	59	38	20
		% FA-complexes	86	84	87	82	86	85
		Calculated labile	6.6	10.7	8.7	24.4	14.0	7.3
		% calculated labile	31	33	30	34	32	32

Appendix

	Time [days]	2.9	3.7	3.9	4.7	4.9	5.7
Mn (-Fe(III))	Dissolved	145	143	118	201	154	171
	Free	96	92	69	152	97	101
	% free	66	65	59	76	63	59
	Inorganic complexes	49	51	49	49	58	70
	% inorganic complexes	34	35	41	24	37	41
	FA-complexes	0.14	0.11	0.08	0.26	0.12	0.09
	% FA-complexes	0.10	0.08	0.07	0.13	0.08	0.05
	Calculated labile	145	143	118	201	154	171
	% calculated labile	100	100	100	100	100	100
Mn (+Fe(III))	Free	96	92	69	152	97	101
	% free	66	65	59	76	63	59
	Inorganic complexes	49	51	49	49	58	70
	% inorganic complexes	34	35	41	24	37	41
	FA-complexes	0.135	0.107	0.078	0.248	0.118	0.084
	% FA-complexes	0.09	0.08	0.07	0.12	0.08	0.05
	Calculated labile	145	143	118	201	154	171
	% calculated labile	100	100	100	100	100	100
Cu (-Fe(III))	Dissolved	110	64	49	49	37	25
	Free	0.095	0.043	0.018	0.029	0.011	0.008
	% free	0.087	0.067	0.038	0.058	0.030	0.033
	Inorganic complexes	5.18	2.54	1.47	0.89	0.76	0.64
	% inorganic complexes	4.7	3.9	3.0	1.8	2.0	2.6
	FA-complexes	104	62	47	48	37	24
	% FA-complexes	95	96	97	98	98	97
	Calculated labile	26.1	14.9	11.0	10.6	8.1	5.5
	% calculated labile	24	23	22	21	22	22
Cu (+Fe(III))	Free	0.448	0.215	0.094	0.174	0.066	0.047
	% free	0.41	0.33	0.19	0.35	0.18	0.19
	Inorganic complexes	24.4	12.7	7.5	5.4	4.4	3.7
	% inorganic complexes	22	20	15	11	12	15
	FA-complexes	85	52	41	44	33	21
	% FA-complexes	77	80	84	89	88	85
	Calculated labile	41.8	23.2	15.9	14.3	11.1	8.0
	% calculated labile	38	36	32	29	30	32

Appendix

	Time [days]	-0.3	0.9	1.2	1.7	1.9	2.7
Pb (-Fe(III))	Dissolved	0.13	0.20	0.18	0.36	0.25	0.18
	Free	0.002	0.002	0.003	0.006	0.002	0.002
	% free	1.7	1.1	1.6	1.8	0.9	1.1
	Inorganic complexes	0.071	0.118	0.077	0.180	0.143	0.098
	% inorganic complexes	54	60	44	50	57	54
	FA-complexes	0.057	0.077	0.096	0.170	0.105	0.082
	% FA-complexes	44	39	54	48	42	45
	Calculated labile	0.08	0.14	0.10	0.22	0.17	0.12
	% calculated labile	65	69	56	62	66	64
Pb (+Fe(III))	Free	0.003	0.003	0.004	0.009	0.003	0.003
	% free	2.3	1.5	2.5	2.5	1.2	1.6
	Inorganic complexes	0.094	0.152	0.119	0.248	0.193	0.138
	% inorganic complexes	72	77	67	70	77	76
	FA-complexes	0.033	0.042	0.053	0.099	0.054	0.041
	% FA-complexes	26	21	30	28	22	23
	Calculated labile	0.10	0.16	0.13	0.28	0.21	0.15
	% calculated labile	80	83	76	78	83	82
Fe	Dissolved	161	219	238	449	335	304
	Free	1.2E-08	1.2E-08	3.2E-08	6.5E-08	1.4E-08	2.4E-08
	% free	7.5E-09	5.3E-09	1.3E-08	1.4E-08	4.1E-09	7.8E-09
	Inorganic complexes	22	60	43	118	112	101
	% inorganic complexes	14	27	18	26	33	33
	FA-complexes	139	159	195	331	223	202
	% FA-complexes	86	73	82	74	67	67
	Calculated labile	50	92	82	184	157	142
	% calculated labile	31	42	34	41	47	47

Appendix

	Time [days]	2.9	3.7	3.9	4.7	4.9	5.7
Pb (-Fe(III))	Dissolved	0.88	0.37	0.20	0.15	0.12	0.03
	Free	0.014	0.005	0.002	0.003	0.001	0.000
	% free	1.6	1.5	1.1	1.8	1.1	1.2
	Inorganic complexes	0.474	0.198	0.107	0.053	0.056	0.019
	% inorganic complexes	54	54	54	35	46	56
	FA-complexes	0.387	0.163	0.091	0.097	0.065	0.015
	% FA-complexes	44	45	45	63	53	43
	Calculated labile	0.57	0.24	0.13	0.08	0.07	0.02
	% calculated labile	65	64	64	49	58	65
Pb (+Fe(III))	Free	0.019	0.008	0.003	0.005	0.002	0.001
	% free	2.2	2.1	1.6	3.0	1.7	1.6
	Inorganic complexes	0.654	0.276	0.150	0.090	0.085	0.027
	% inorganic complexes	75	75	75	59	70	77
	FA-complexes	0.202	0.083	0.047	0.058	0.035	0.008
	% FA-complexes	23	23	24	38	29	22
	Calculated labile	0.71	0.30	0.16	0.11	0.09	0.03
	% calculated labile	82	82	81	70	77	83
Fe	Dissolved	743	475	330	326	280	202
	Free	7.7E-08	5.0E-08	2.0E-08	5.2E-08	1.7E-08	1.3E-08
	% free	1.0E-08	1.1E-08	5.9E-09	1.6E-08	6.1E-09	6.5E-09
	Inorganic complexes	225	158	106	67	78	59
	% inorganic complexes	30	33	32	20	28	29
	FA-complexes	518	317	224	259	203	143
	% FA-complexes	70	67	68	80	72	71
	Calculated labile	329	221	151	119	118	88
	% calculated labile	44	47	46	36	42	43

Appendix

Metal content in periphyton (S2)

Content is referred to as [nmol/g dw] for Cd, Cu, Pb and as [µmol/g dw] for Zn, Mn and Fe.

	Time [days]	-0.3	1.2	1.7	1.9	2.7	2.9
Cd	Total	0.62 ± 0.02	2.4 ± 0.3	2.6 ± 0.2	2.5 ± 0.7	1.7 ± 0.2	2.8 ± 0.7
	Intracellular	0.42 ± 0.22	1.9 ± 0.2	1.9 ± 0.5	2.0 ± 0.6	1.5 ± 0.0	1.5 ± 0.4
	Adsorbed	0.20	0.44	0.68	0.45	0.27	1.3
	% adsorbed	32	19	26	18	16	46
Zn	Total	1.1 ± 0.2	2.7 ± 0.2	3.0 ± 0.2	2.3 ± 0.4	1.8 ± 0.5	3.6 ± 1.1
	Intracellular	0.69 ± 0.24	1.8 ± 0.5	1.9 ± 0.6	1.6 ± 0.2	1.6 ± 0.3	1.1 ± 0.3
	Adsorbed	0.40	0.93	1.1	0.67	0.20	2.5
	% adsorbed	37	34	35	29	11	69
Mn	Total	2.7 ± 0.5	9.6 ± 0.9	9.7 ± 0.5	8.3 ± 1.6	8.9 ± 2.5	18 ± 6
	Intracellular	2.4 ± 1.1	8.2 ± 2.4	9.5 ± 3.0	8.5 ± 1.0	11 ± 3	9.9 ± 2.3
	Adsorbed	0.23	1.4	0.22	-0.18	-2.3	8.4
	% adsorbed	9	15	2	-2	-26	46
Cu	Total	66 ± 17	488 ± 32	695 ± 87	488 ± 73	354 ± 78	898 ± 317
	Intracellular	42 ± 15	390 ± 125	570 ± 165	433 ± 44	368 ± 85	336 ± 76
	Adsorbed	24	98	125	54	-14	562
	% adsorbed	37	20	18	11	-4	63
Pb	Total	10 ± 2	79 ± 7	107 ± 8	73 ± 10	51 ± 14	106 ± 35
	Intracellular	8.9 ± 5.3	56 ± 18	76 ± 25	52 ± 7	52 ± 13	37 ± 8
	Adsorbed	1.2	23	32	21	-1.3	69
	% adsorbed	12	29	30	28	-3	65
Fe	Total	7.4 ± 1.5	61 ± 6	87 ± 5	52 ± 10	44 ± 11	86 ± 28
	Intracellular	7.4 ± 3.4	62 ± 18	91 ± 29	54 ± 7	60 ± 16	50 ± 12
	Adsorbed	0.06	-0.82	-4.0	-1.9	-16	36
	% adsorbed	1	-1	-5	-4	-36	42

Appendix

	Time [days]	3.7	3.9	4.7	4.9	5.7
Cd	Total	2.5 ± 0.2	1.7 ± 0.2	2.4 ± 0.0	2.7 ± 0.4	2.5 ± 0.4
	Intracellular	2 ± 0.1	1.3 ± 0.2	1.9 ± 0.3	1.7 ± 0.2	1.9 ± 0.3
	Adsorbed	0.47	0.37	0.55	1.0	0.62
	% adsorbed	19	22	23	38	25
Zn	Total	3.6 ± 0.4	1.5 ± 0.2	3.4 ± 0.4	4.0 ± 0.5	2.7 ± 0.3
	Intracellular	1.8 ± 0.4	1.2 ± 0.5	2.0 ± 0.3	2.1 ± 0.1	1.6 ± 0.6
	Adsorbed	1.8	0.28	1.4	2.0	1.1
	% adsorbed	50	18	41	49	40
Mn	Total	16 ± 2	6.4 ± 0.9	17 ± 3	21 ± 4	18 ± 2
	Intracellular	14 ± 4	9.2 ± 4.0	20 ± 3	20 ± 2	17 ± 5
	Adsorbed	2.3	-2.8	-2.5	1.8	0.77
	% adsorbed	14	-44	-15	8	4
Cu	Total	721 ± 110	289 ± 38	731 ± 90	865 ± 145	573 ± 48
	Intracellular	536 ± 169	328 ± 133	595 ± 98	592 ± 26	411 ± 112
	Adsorbed	186	-39	136	273	161
	% adsorbed	26	-14	19	32	28
Pb	Total	114 ± 14	34 ± 5	82 ± 13	112 ± 18	79 ± 13
	Intracellular	64 ± 18	38 ± 18	68 ± 13	75 ± 5	63 ± 19
	Adsorbed	50	-3.3	14	37	15
	% adsorbed	43	-10	17	33	20
Fe	Total	82 ± 8	31 ± 4	70 ± 11	91 ± 12	69 ± 9
	Intracellular	74 ± 21	48 ± 21	82 ± 14	89 ± 8	70 ± 19
	Adsorbed	8.2	-17	-13	2.4	-1.4
	% adsorbed	10	-54	-18	3	-2

Appendix

Pictures

Kinetics of cadmium accumulation in periphyton under freshwater conditions

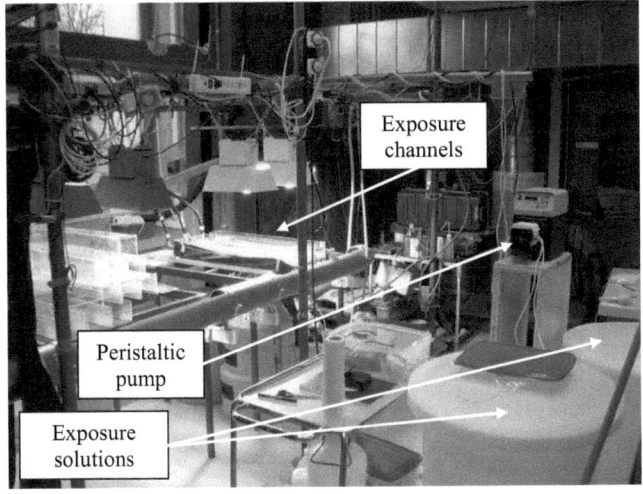

Experimental design

Exposure channels

Close-up of periphyton slides

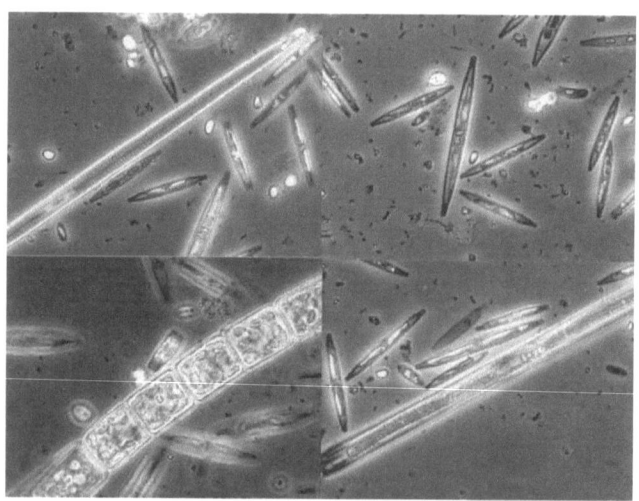

Microscope pictures of periphyton

Appendix

Accumulation of cadmium in periphyton under various freshwater speciation conditions

Experimental design

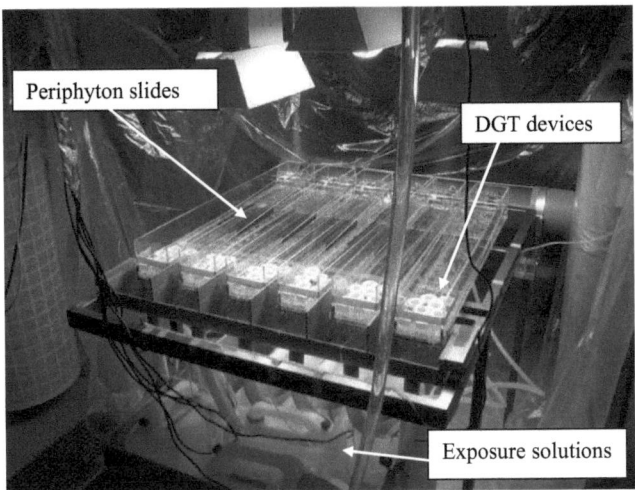

Exposure channels

Appendix

Exposure solutions

Close-up of periphyton slides

Appendix

Cadmium speciation and accumulation in periphyton in a small stream during rain events

Field site

Teflon racks with periphyton slides

Appendix

Teflon racks with periphyton slides and DGT devices

Appendix

Periphyton slides

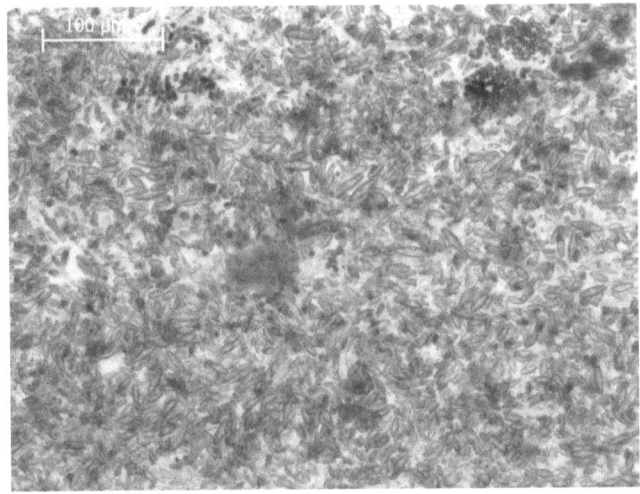

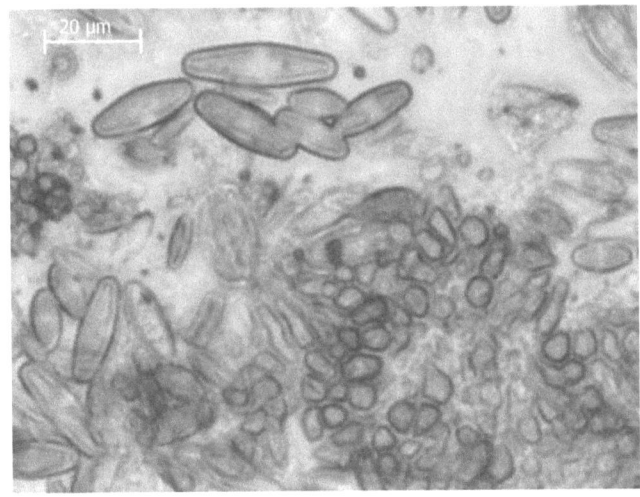

Acknowledgements

I look back on four exciting, challenging, good but also sometimes hard and exhausting years at Eawag. I am grateful that I had the opportunity to gain insight into various fields of environmental toxicology, the opportunity to take part in congresses, to improve my presentation skills and to meet a lot of great and friendly people.

I want to thank many people, who supported me during my whole Ph.D. thesis not only with their knowledge and help but also emotionally. First of all I want to thank my two supervisors Laura Sigg and Renata Behra who gave me the opportunity for this diversified and miscellaneous thesis and introduced me into the various fields about metals, algae and periphyton. My ex-girlfriend Afra who supported me during my whole thesis with her love and kindness. David Kistler for his help with the HR-ICP-MS, in the laboratory and during the field study. Adrian Ammann for his help with the HR-ICP-MS, data processing and interpretation. Bettina Wagner for her help in the laboratory, the "Versuchshalle" and the field. Enrique Navarro for setting up the whole channel system, his introduction, explanations, help during the experiments, data interpretation and discussions. Niksa Odzak for his help making the DGT devices, understanding the principles and calculations. Peter Gäumann and his team for their advice and help in the construction of equipment for the channels and field. Madeleine Langmeier, Richard Illi and their team for the analysis of water and periphyton samples. Sebastien Meylan and Jacqueline Traber for the analysis of natural organic matter and the discussion of the results. Esther Keller for the determination of species composition. All other colleagues from Eawag for their help and discussions. And finally the Swiss National Foundation (SNF) for their financial support.

I want morebooks!

Buy your books fast and straightforward online - at one of the world's fastest growing online book stores! Environmentally sound due to Print-on-Demand technologies.

Buy your books online at
www.get-morebooks.com

Kaufen Sie Ihre Bücher schnell und unkompliziert online – auf einer der am schnellsten wachsenden Buchhandelsplattformen weltweit!
Dank Print-On-Demand umwelt- und ressourcenschonend produziert.

Bücher schneller online kaufen
www.morebooks.de

OmniScriptum Marketing DEU GmbH
Heinrich-Böcking-Str. 6-8
D - 66121 Saarbrücken
Telefax: +49 681 93 81 567-9

info@omniscriptum.com
www.omniscriptum.com

Printed by Books on Demand GmbH, Norderstedt / Germany